D0648805

THE SLOAN TECHNOLOGY SERIES

Dark Sun: The Making of the Hydrogen Bomb by Richard Rhodes

Dream Reaper: The Story of an Old-Fashioned Inventor in the High-Stakes World of Modern Agriculture by Craig Canine

Turbulent Skies: The History of Commercial Aviation by Thomas A. Heppenheimer

Tube: The Invention of Television by David E. Fisher and Marshall Jon Fisher

The Invention That Changed the World: How a Small Group of Radar Pioneers Won the Second World War and Launched a Technological Revolution by Robert Buderi

Computer: A History of the Information Machine by Martin Campbell-Kelly and William Aspray

Naked to the Bone: Medical Imaging in the Twentieth Century by Bettyann Kevles

A Commotion in the Blood: A Century of Using the Immune System to Battle Cancer and Other Diseases by Stephen S. Hall

Beyond Engineering: How Society Shapes Technology by Robert Pool

The One Best Way: Frederick Winslow Taylor and the Enigma of Efficiency by Robert Kanigel

Crystal Fire: The Birth of the Information Age by Michael Riordan and Lillian Hoddesen

Insisting on the Impossible: The Life of Edwin Land by Victor K. McElheny

Silicon Sky: How One Small Start-Up Went Over the Top to Beat the Big Boys Into Satellite Heaven by Gary Dorsey

City of Light: The Story of Fiber Optics by Jeff Hecht

Visions of Technology: A Century of Provocative Readings edited by Richard Rhodes (forthcoming)

Silicon Sky

Other books by Gary Dorsey:

Congregation: The Journey Back to Church

The Fullness of Wings: The Making of a New Daedalus

Silicon Sky

How One Small Start-Up Went Over the Top to Beat the Big Boys into Satellite Heaven

Gary Dorsey

A CORNELIA AND MICHAEL BESSIE BOOK

PERSEUS BOOKS

Reading, Massachusetts

Many of the designations used by manufacturers and sellers to distinguish their products are claimed as trademarks. Where those designations appear in this book and Perseus Books was aware of a trademark claim, the designations have been printed in initial capital letters.

ISBN 0-7382-0094-8
Library of Congress Catalog Card Number: 99-60039

Perseus Books is a member of the Perseus Books Group

A Cornelia and Michael Bessie Book

Jacket design by Bruce W. Bond
Text design by Greta D. Sibley & Associates
Set in 11-point Minion by Vicki L. Hochstedler

123456789—0302010099
First printing, March 1999

Find us on the World Wide Web at
http://www.aw.com/gb/

For
Hugh Dodd Dorsey, Engineer

I can take down your fences, lead you back to the garden.

—Thomas Pynchon,
Gravity's Rainbow

Contents

Preface to the Sloan Technology Series

Technology is the application of science, engineering, and industrial organization to create a human-built world. It has led, in developed nations, to a standard of living inconceivable a hundred years ago. The process, however, is not free of stress; by its very nature, technology brings change in society and undermines convention. It affects virtually every aspect of human endeavor: private and public institutions, economic systems, communications networks, political structures, international affiliations, the organization of societies, and the condition of human lives. The effects are not one-way; just as technology changes society, so too do societal structures, attitudes, and mores affect technology. But perhaps because technology is so rapidly and completely assimilated, the profound interplay of technology and other social endeavors in modern history has not been sufficiently recognized.

The Sloan Foundation has had a long-standing interest in deepening public understanding about modern technology, its origins, and its impact on our lives. The Sloan Technology Series, of which the present volume is a part, seeks to present to the general reader the stories of the development of critical twentieth-century technologies. The aim of the series is to convey both the technical and human dimensions of the subject: the invention and effort entailed in devising the technologies and the comforts and stresses they have introduced into contemporary life. As the century draws to an end, it is hoped that the series will disclose a past that might provide perspective on the present and inform the future.

The Foundation has been guided in its development of the Sloan Technology Series by a distinguished advisory committee. We express deep gratitude to John Armstrong, Simon Michael Bessie, Samuel Y. Gibbon, Thomas P. Hughes, Victor McElheny, Robert K. Merton, Elting E. Morison (deceased), and Richard Rhodes. The Foundation has been represented on the committee by Ralph E. Gomory, Arthur L. Singer, Jr., Hirsh G. Cohen, and Doron Weber.

—Alfred P. Sloan Foundation

Preface

The Space Age is dead, they say, and commerce is King.

Childhood fantasies once sparked by model rocket contests, science fiction adventures, and romantic memories of barnstormers on the moon still find fulfillment in a new spaceflight revolution, though not quite in the same way that the heroic deeds of cowboy astronauts made breathtakingly possible thirty years ago. In fact, the near heavens now have been so well defined, so thoroughly charted and domesticated, that they may no longer be accurately called a frontier at all.

But there is a Klondike overhead, and cowboys of a different sort cruise the skies.

Orbital stations and artificial constellations today frame dark industrial settings for billion-dollar shoot-outs among mavericks and aerospace superpowers. At the end of this year, corporations around the world will have poured $50 billion into a ferocious battle to build, launch, and sell a broad range of satellite services to massive consumer markets around the globe. Commercial satellites will eventually make Dick Tracy–style wristwatch radio communicators available to almost anyone on Earth. Satellites will identify potholes on highways, locate lost backpackers in the Appalachians, relay a cornucopia of TV channels to the Amazon, and provide high-speed, broadband Internet services for intrepid travelers wherever they roam.

In only the last few years, a fundamental shift has occurred. The historic role of government in space ventures now pales compared with efflorescent commercial businesses springing quickly into place. Satellite trajectories from low-Earth orbit, about 450 miles high, to geosynchronous platforms 22,000 miles away, are creating high-tech infrastructures for a monumental industry that no longer relies on Defense Department contracts and pork-barrel politics. In 1997, for the first time in history, the number of commercial satellite launches nearly equaled the number of new government and military satellites in space. The investment of private capital in new launch systems overshadowed money spent for similar government projects. The number of in-orbit, active commercial satellites nearly doubled the number of active government satellites.

The profound realignment of capital that occurred in 1997 has been building rapidly since the end of the cold war, and the trend is not likely to reverse. The commercial space industry grows at an exponential rate. In some ways the transition actually started after the Persian Gulf War, with an explosion of interest in the Global Positioning System and the sudden boom in direct-to-home satellite television products. With the advent of satellite-positioning devices for consumers and new global markets suddenly opening up for satellite television at home and abroad, enormous sums of investment capital came flooding into what was once a fairly tepid commercial space industry. From 1995 to 1997, a half-billion dollars a month in new private capital were invested in commercial satellite projects alone.

Innovative ideas for space-based services, which incubated for at least a decade because of a lack of start-up funding, suddenly came to life, and the effects are visible. Small umbrella-like satellite antennas have mushroomed across rooftops in cities across the world. Thousands of boaters now purchase GPS receivers at less than $100 a pop to chart their paths across seas. Gas station operators rely on satellite systems for credit card verification so an entire transaction may be handled directly at the pump, relieving customers from having to run through the rain or stand in line at a checkout counter.

But most significantly, dozens of corporations—from little start-up entrepreneurs like Orbital Sciences to giants led by dominating figures like Bill Gates and Craig McCaw—have designed the most audacious orbiting networks to link individuals wirelessly from any place, at any time, to anyone else in the world. Over the next few years, these businesses will launch and place as many as one hundred satellites a year into orbital positions, creating a swarm of new products and services for the consumer marketplace. The benefits of space—for the first time—will come down to Earth.

In 1991, when members of the Alfred P. Sloan Foundation invited me to write for their twentieth-century technology series, it was hard to think of "commercial space" as anything but an oxymoron. For twenty years the people who talked about creating new businesses tended to stand out on the fringes, and many of them predicted, erroneously, that the revolution, when it came, would begin with space tourism or electrical power generation or the mining of asteroids or the creation of lifesaving, cancer-curing pharmaceuticals from a space station. The products and services that now lead the revolution—communications, navigation, and remote imaging—rounded out the second tier of predictions, at best.

I was lucky, though. I chose to write about one particularly exciting project in one particularly prescient company. This book tells their story, an adventure of small groups of people simultaneously engaged in creating a new technology, building a business plan, identifying markets, and growing a corporation with a bold vision for expanding the boundaries of private enterprise in space. What I have written is, in that respect, only one window into the new world of space commerce, but it is also a reflection of larger efforts in a revolution taking place in new orbits literally around the globe.

Thanks are in order.

I am particularly grateful to David Thompson, chief executive officer of the Orbital Sciences Corporation, for opening the door and standing aside so I could observe every level of his business. He invited me, without reservation, to attend all meetings, question any engineer, and wander wherever I desired. And so I did, with tape recorder, notepad and pen, quietly invading the corridors and conference rooms of the Orbital Sciences Corporation. Without intrusion, I witnessed the remarkable development of a new technology—and in small part, a new industry—from its earliest stages of design in 1992 to its launch in 1995. The complex, difficult, and sometimes controversial nature of that enterprise, which played out dramatically in the lives of Orbital's engineers and executives, might have forced a less confident person to limit my access or to demand revisions in the final manuscript to cloak awkward moments of indecision or miscalculation, so the book would reflect more favorably on the business and its leadership. I am happy to report that Thompson considered interference of any sort, at every step of the process, simply out of the question. That, I believe, is a remarkable measure of his professionalism and his commitment to the Sloan's purpose in educating the public about the realities of business and technology.

To Alan Parker, David Schoen, Dave Steffy, and Jan King, who led the engineering and marketing teams, I also am indebted for making me welcome and, in some cases, offering to tutor me in important aspects of marketing, new communications technology, and aerospace engineering. Each man is blessed with a pioneering spirit, and I can honestly say that it has been a privilege to be their witness.

Veteran engineers, wolf pups, Freshouts, satellite prima donnas, and grizzly technicians who populated the project made me feel like part of the team over our three years together. Although as a journalist I manage to stay at least a step removed from my subjects, the task was made considerably easier by a team that respected my distance and also encouraged a relationship that brought me into their lives whenever I needed. It was they who explained esoteric concepts of orbital sciences to me, found me a place at the dinner table on their longest days, and risked sharing private thoughts and conversations that could have imperiled their jobs.

As far as I know, this book is the first to document, from the inside, the story of an aerospace project from design to launch, and after seven years I can see why. This has been the longest single writing project of my life. The ups and downs of such prolonged work, the capricious and sometimes malicious nature of book publishing during this period, and the financial difficulties that it entailed have lent an even greater appreciation for the corporate struggles of entrepreneurial ventures like Orbital's. In many ways, all entrepreneurs depend on the grace of angels, and, thankfully, I have had my share.

So I also offer my gratitude to Art Singer, Richard Rhodes, and Vic McElheny, each of whom contributed to another launch for Sloan's excellent book series and, in separate but significant ways, encouraged my vision for this project from start to finish. Most especially, I want to acknowledge the support and wisdom of an editor who may be the last true gentleman in publishing, Michael Bessie, who never once said to me "ship and shoot," while he patiently and single-handedly cleared the path for publication, and to Amanda Cook at Perseus Books, who brought it all home.

Members of my family, John and Mary Ann Dorsey, graciously took me in and gave me a place to stay when funds ran low, and my wife, Jan Winburn, managed to endure months of separation with a kind of understanding only the best wives or editors manage. My love and gratitude go to them, as always.

More than forty years after the launch of Sputnik, as I recount the stories of young engineers, some of the first satellites still spin in orbit. Fuzzy little satellites—humorous instruments to look at today: glazed over with solar cells, guts wrapped in foil, fitted with gangly limbs and protruding, insectlike antennae and eyes. Here is a technology that came into the world a symbol as fearsome as a mushroom cloud, and over a period of a few decades became a quiet, quaint, silently accepted part of daily life, as much a part of our world as an iron bridge, a personal computer, or an electric toaster.

The fact that engineers have come to call them "birds" is no nearsighted mistake. An affectionate stretch of the imagination makes these thrillingly active machines almost visible above the trees, uniting the world in fantastic ways. Watching the changing face of the Earth, coding its atmosphere, relaying messages from global village to global village, satellites provide a reassuring constancy that one might only pray for from other less beastly-looking celestial creatures.

It has taken a long time to realize their full potential, but in the next few years, as they do, the world they changed will feel their effect again. And we will never be the same.

Bulletin Board

"Who says a satellite network can't be a mom-and-pop operation? Hey, not so long ago, no one at IBM or Digital thought there was a market for a personal computer."

—Heather Miller
Wired 4.09

"There's only one outstanding example of a good return on government investments in space: communication satellites. We've put $500 billion into the civil space program, and today we get $4 billion a year out of it. We would have done better if we put our money in a savings account."

—Retired aerospace executive
Interview with author
January 1992

"Though we can be quite smug about inventing the use of the small nuclear team for small satellite development, it is the approach primitive humans probably used to kill a large animal a million years ago. One particularly experienced, capable or aggressive person serves as a focus, but the activities of the small group—consisting of spear and hemp carriers, scout, runner, and maybe one gifted in heavy lifting—were organized through the clear communication of a common goal (e.g., to kill while avoiding being killed). They frequently cooperated through direct communication, and we can be sure this communication was not characterized by too many ICDs and CDRLs (Interface Control Documents and Contract Data Requirement Lists)."

—Rick Fleeter
"Management of Small
Satellite Programs"
September 2, 1992

"The life cycle of a technology is sometimes conceived as a linear process, from invention or research to development to production and diffusion. In fact, the evolution of most technologies is neither simple nor linear."

—Pamela Mack
Viewing the Earth

CHAPTER ONE

January 1992

Microsoft in Microspace

Canada geese rose over a thin strand of pine trees, eased out of formation, and glided quietly down into a clear, cold, man-made pond. A frigid, windy afternoon, the day had brought word of fourth-quarter profits, new business, and a return to the heavens. From the bright panorama of David Thompson's second-floor wall of windows, I watched the impressive gaggle gather itself and again take flight, joining pluming trails of jetliners and clusters of gossamer clouds crossing our line of sight.

From the height where he worked, the wintry sky appeared gray and glossy with a sheen like Polarized glass. A few months before the first full crew of Freshouts joined the company, the chief executive officer could not point to the position of a single working satellite bearing the name Orbital. He had no spacecraft control center at his command, few engineers in his labs.

The sky was merely a sketch pad; the bright canopy beyond the window, imaginary terrain.

"Look," he said.

The first satellite broke from the east across ragged treetops and rose like a moon. It vanished. Another hurtled from the north, sped upward, just as a third streaked overhead, crisscrossed paths with a fourth, then disappeared. Ten more flew by. Fourteen. Twenty-six.

I tried to imagine it for myself—twenty-six satellites scratching paths across the skyline.

For more than five years, Thompson had planned an intricately crafted webwork almost within sight, a constellation of Earth-swarming satellites spread in symmetrical patterns a few hundred miles above the Earth, coursing between 5 and 40 degrees above the horizon. What he saw, standing at a second-floor window in a pastoral office park in northern Virginia, was not quite clear to me yet, though I listened and watched curiously as he sectioned off pieces of the heavens like a squatter.

It was mid-January 1992, a Friday, when we first met—"casual day," according to a message board at the corporation's headquarters in Fairfax County. Preliminary financial reports, marked "confidential," made their way around offices on the second floor. The first raw accounting figures indicated the little company had invested 25 percent of its income in R&D projects during 1991; sales hovered neatly around $130 million; employee productivity was still about 50 percent higher than at any other major company in the space business. For the first time in three years, the Orbital Sciences Corporation could expect to bank a profit—*two percent*—after taxes.

Someone had scheduled a pizza party to celebrate the good news in the high bay at the Sullyfield Office Park, a few miles north where the company's space division conducted business. At Orbital executives believed that occasional pizza parties, like the regular availability of free sodas, made a statement about the kind of grimjack, over-the-top culture they wanted to maintain even as the business grew—something like the playfully intense Microsoft cynosure but on a fresh plot of competitive turf in the new celestial frontier, the emerging domain they liked to call Microspace. Division chief Bob Lovell, who had come to lead the technical teams after a long and distinguished career at NASA, insisted that he would be at the party, as did Antonio Elias, the MIT rocket scientist who had recently met President George Bush to accept the National Medal of Technology for the company's remarkable new Pegasus rocket.

They had little regard for formal celebrations or rituals of pomp and circumstance. But there was no question that while Antonio and Bob would have to reorganize their entire afternoons and fight their way through technical questions and contractual nits springing up over the company's six new development projects, they would never miss a chance to share a cold soda and a slice of pepperoni pizza with the young crews.

Thompson, on the other hand, had been working the phones all morning from his executive office, and from there his face would catch the last angles of January sunlight as the sky turned orange, lighting the long glass walls of his office like a prism, illuminating globes and world maps that surrounded him. Dressed all in white like a twenty-first-century Lord Jim, he had excused himself from social obligations for the day, explaining simply, "I have to raise $100 million."

His face yielded no sign of irony or boastfulness, though he did fold his hand across his stomach while he talked, like a man gripping his guts.

One hundred million was chump change for some aerospace companies, though Thompson's people desperately needed at least that from a multinational corporate sweetheart who had not yet been identified. Although it was a relatively piddling sum elsewhere, without $100 million his Orbcomm project would be doomed even before the first bit flipped.

Motorola, for instance, budgeted more than $3 *billion* to launch Iridium, its satellite constellation telephony service, in 1998. Although Orbital expected someday to evolve a global satellite voice system from its less showy, data-only Orbcomm constellation—the "bongo drum" version, Thompson called it—the most important goal was to beat everyone else to market. A three-bit player could afford neither to wait so long nor to cost as much as the giants. The great challenge facing Thompson was not simply luring investors, raising money, and building a revolutionary flotilla of miniature communication satellites. He had to fulfill the preposterous claim that tiny Orbital Sciences could bring a global telecommunications service to the world market by 1994 on a ridiculously meager budget, and that he could do it first.

It is important to note that David Thompson was not, you see, gazing out his floor-to-ceiling windows watching Canada geese fly, as I did when I first arrived. He was not some wistful, ponderous geek sitting at a big desk, nervously pecking away at a Gateway 2000.

Around the second floor, his engineers spoke in hot strings of acronyms—PCS, ISDN, GPS, SGI, CDMA—and his small marketing team met over business plans that shirked Manhattan, Chicago, Miami, and L.A., in favor of the faraway pampas, the backstreets of Caracas, the broad landscapes around Alice Springs and Phnom Penh, and the dusty streets of Kampala and Tombouctou. Within three to five years, it had been reported in trade journals, Orbital's small teams would launch a constellation of twenty-six small satellites around the Earth and begin selling a kind of *Star Trek* global communicator for less than $200 a pop on international markets. Giant competitors like Motorola, Lockheed, Hughes, Loral, and TRW, Thompson had predicted, would choke on their dust.

However, as we talked I kept thinking that he was not the imposing figure suggested by great ambitions. He did not look like a person bred to conquer mastodons. Thompson was actually a pleasant ectomorph, a relatively small man, still in his mid-thirties, as modest and beguiling as a young seminarian.

Wearing white cinch-waist jeans, a white polo shirt, and white leather deck shoes, he appeared to be every inch the New Age entrepreneur of space, a smaller, more elegant version of the ham-fisted DARPA colonels who ruled over bottomless black budgets that fueled the nation's most mysterious and sophisticated space missions.

His brown hair, stylishly thick and irregular, swooped across his forehead, played over his ears, and frothed up around his collar. I had read one newspaper account that said he resembled Harrison Ford, the actor. But in this casual getup, he looked more like a rich boy on vacation, as svelte and buoyant as top seed in a collegiate tennis match, a choice date for a debutante. The son of a middle-class science teacher from South Carolina, he carried a surprising air of wealth and good breeding that suggested something rather different from his pedestrian origins—a famous New England heritage, perhaps, or a family line that could be traced to the founding of Charleston.

As thin as a sapling, energetic, and handsome, his greatest assets were less physical than, as was often noted, the most potent and abstract sort: a strong will; a wealth of intelligence; a mighty ambition. At the cusp of mid-career, he told friends he had only begun to address new ideas that would make space as profitable as Arabian oil fields.

As he talked, I had to keep reminding myself that we were meeting only in rented office space. The tenants next door were small real estate firms. His company had only turned a profit after taxes for the first time. I also had to remember that in the business of satellite commerce, particularly among those companies racing to settle the low-Earth frontier, everyone stank of hype.

Perhaps it was unavoidable. At a time when the industry needed young role models, awards and honors came to him regularly. They had made him, in a sense, a premature hero:

- *He had recently been named "satellite executive of the year" by members of the industry, even though his company had yet to launch its first successful satellite.*
- *He had just been identified as "a potential billionaire of the 21st century" by* Inc., *a magazine for entrepreneurs, even though he and his pals were only a few years away from the time they had kept the company afloat on three maxed-out credit cards.*
- *Harvard Business School had selected the company for its third case study and invited Orbital executives to talk to MBA classes, even though the business lacked money to staff more than a dozen engineers for its new Orbcomm satellite project.*

- *Influential insiders around the Washington Beltway spoke of him as next in line to take over as deputy administrator of the troubled NASA space agency, even though he often said publicly that he resisted the thought of ever becoming part of the NASA establishment.*
- *Even though he complained that he wanted nothing more than to break the industry's incestuous relationship with the United States government and prove that the power of private capital could invigorate the aerospace business, fully 90 percent of his company's contracts were signed with NASA, the Department of Defense, and the U.S. Air Force.*

But the most important fact was this: While his example did inspire well-wishers and the hopes of many that the space business could be successfully privatized and shaped by new players, the Orbital Sciences Corporation had yet to demonstrate its substance on the larger and more significant scales of time and profits. His ultimate goal, I had heard—to build nothing less than the world's greatest space company—not only grated against the expectations of some of his employees, but his own sense of heroic mission sometimes appeared to mask a shallow arrogance or subtle gimmickry that had characterized the aerospace industry since its earliest years. In fact, one NASA executive, who later became a booster and enthusiast, remembered meeting Thompson at a briefing when the company was first formed and, after listening for only a few minutes, suddenly realized, "This fellow doesn't have a pot to piss in."

On the other hand, many people recommended him. According to those in the business who knew him best, Thompson was potentially as remarkable in the field of aerospace as Steven Jobs in the business of personal computing. Supporters and admirers included former Secretary of the Air Force John L. McLucas, Apollo astronaut Harrison Schmitt, Paul Kinloch, managing director for investment banking at Lehman Brothers, and Jack Kerrebrock and Lennard Fisk, both onetime associate administrators of NASA. In just a few years, Thompson and company had earned the respect of some of the most honored people in the business, a distinguished variety of astronauts, NASA administrators, professors, industrialists, and politicians, some of whom sat on his board of directors. Many of them took him seriously enough that they chose to gamble personally—that is, financially—on his schemes.

The truth was, Thompson had only recently emerged in the industry, but he was presently considered—and celebrated as—the world's first true entrepreneur of space. Indeed, he was the reigning wunderkind of the business.

Although several corporations had announced intentions to settle the new frontier of low-Earth orbit, the only complete constellations in existence at the time amounted to those comprising the natural firmament. Because he and his colleagues had planned the project for several years—even to the point where they had designed and built their own rocket to launch the satellites—and not even the largest companies preceded him, his corporation was dead on course to be among the first pioneers to domesticate the promising domain of near space. The first to develop those dark but proximate orbits—located between 400 and 1,200 miles above the Earth's surface—would likely strike a major claim in what would, in a decade, become a twenty-first century gold rush, the first race to space with commercial, rather than political, motives.

As we talked, it also became clear that he was not a person who worried too much about the odds.

Winter sunlight danced against the tinted glass of his pie-wedged office and constantly drew my attention back to the horizon outside. I listened as he set corner stakes in the most sublime of territories: a solitary cumulus cloud to the east; a few miles south where the contrails of jets lingered; a distant smudge of smog floating over Interstate 66.

Pierce them, then it was all skyward, only as distant as the interstate highway linking Washington to Boston. Straight up. Blue turf. Open fields. High cotton.

Well spoken and witty, enthusiastic and affable, he talked in a stream during the few last hours of a cold afternoon, reviewing his tactics and strategies with a sense of boyish delight.

"Crazy?" he repeated when I, at one point, expressed the obvious doubts. Then quickly, without hesitation: "Well, yes, it is pretty dramatic insanity around here."

Then suddenly laughing, he only mocked my question with his own: "The only problem I have at the moment is, where the heck am I going to get $100 million?"

In 1992, when I first began to explore the unlikely business of commercial space, it was clear, from Washington to Los Angeles, that the space age was quickly receding in the rearview mirror of the twentieth century. A parody of formerly heroic times, the withering cult of aerospace still held the interest of a bemused public mostly because of its campy, bloated vestige of cold war ambitions and the awkward efforts of the federal government and private industry to revive something of the Apollo spirit before congressional budget-cutters clipped its wings.

Exactly when the decline set in was hard to say, but the image of Christa McAuliffe's parents in 1986 standing beneath the strange, swirling smoke of the *Challenger* explosion came most clearly to mind. The loss of the shuttle marked the first in a strange series of very public disasters and malfunctions that suggested what rather quickly became identified as a far-reaching malaise. After humiliating, public investigations of the *Challenger* disaster, in 1990, the billion-dollar Hubble Telescope turned its bleary eye on the universe and aroused another wrenching public inquiry. A couple of years after that, TV networks broadcast live coverage of American astronauts on another billion-dollar shuttle mission fumbling hour after hour as they tried, unsuccessfully, to save a misguided Intelsat VI with what amounted to a million-dollar device that looked like nothing more than a deluxe tire iron. Cynicism and incompetence had settled into the heart of the nation's space program.

In response, the respected aerospace executive Norm Augustine issued stinging, sometimes satiric attacks against a strangled bureaucratic system that had overtaken the soul of his industry. Stories of infamous $17,000 hammers and $60,000 toilet seats became commonplace. In 1989 a pudgy Apollo-era Neanderthal named Maurice Minnifield showed up as a character on TV's *Northern Exposure* wearing a NASA cap that, each week, mocked the agency, the aerospace business, and their foolish, fading glory. The space shuttle, which was by then a lingering example of a 1970s technology frozen in time, experienced one delayed mission after another, and for a few years the American space effort was little more than a terrestrial event, stalled indefinitely in the past.

In 1990 MIT historian Pamela Mack published a well-documented, devastating account of the history of the Landsat program that demonstrated how pork-barrel politics, bureaucratic power blocs, and interest group wrangling had ensnared government satellite programs in communications, Earth observation, and weather forecasting for decades. Since the birth of satellite technology in the 1950s, she noted, "science and technology would support the politics, not the other way around."

At Duke University aerospace historian Alex Roland took aim at the industry's emphasis on gargantuan missions, overwrought technology, and excessive budgets. He compared NASA's manned space program to the pyramids—not just like tombs for pharaohs, but *overengineered* tombs for pharaohs, civic monuments that spoke of corporate gluttony and conspicuous consumption.

"The space community is gathered around NASA now the way a family gathers around a wealthy, powerful, old, and dysfunctional relative," Roland wrote in a 1991 essay. Recalling that NASA had once been driven by mythic concepts, Roland attacked agency administrators as "a bunch of old men who had their peak experiences together as youngsters" and still wanted to relive those days without knowing how to adapt to changes in time and culture. "Explorers," he concluded, "don't age well. We are in a state of suspended adolescence, deferring mature exploitation of space in a childish infatuation with circus."

As bloated corporations underwent scrutiny and downsizing, the industry also became the butt of jokes. In 1992 noted aerospace author William E. Burrows stood before a venerable crowd of industry executives at the annual National Space Symposium and ballyhooed their predicament. Burrows observed that *The New York Times*'s Pulitzer prize–winning space reporter, John Noble Wilford, was now writing more about dinosaurs, paleontology, and "other extinct creatures" than about space missions. "Covering the space community for more than twenty years was perfect preparation," he smirked.

By the time David Thompson announced plans to launch the world's first commercial satellite constellation, the industry was, if not lost in space, certainly floundering. Its greatest achievements were regarded in some places with as much nostalgia as Richard Nixon in China or Marshall McLuhan in polyester, with as much comic distance as cold war iconography, psychological warfare, Tang, space dogs in orbit, and monkeys fitted with silver Mylar-and-Velcro suits. The hip world's underground communicated in those days not via spacecraft but through a vast interconnected system of cables and wired networks that the new American vice president would call "the information superhighway," and the wider world would soon recognize as the awesome Internet. That satellites might someday take the Net and lift it skyward, making it mobile as well as global, had entered the minds of only the most prophetic or most peculiar savants in aerospace, and for a few years the already-aged Future of the Industry continued to take one pratfall after another, collapsing upon the Present, until the mere mention of Space Age anything—*design, materials, plastic*—evoked only a chuckle among a new generation of engineers born of the Intel Age, who regarded the "final" frontier as a joke, while they touted constant advances in computers resting on desktops as "insanely great."

It may have only been a reflection of how dire times had become that a young man like David Thompson created a stir when he started comparing satellites to personal computers and coined the word Microspace to describe his vision of how a small entrepreneur might find new ways to "do" space in the 1990s. It was not entirely obvious how a constellation of small satellites could serve massive consumer markets the way the Apple II computer did, but in any case his ideas began to attract a following among people who were, if not immediately persuaded, at least willing to indulge his passionate desire to blast old paradigms.

In a keynote address in 1989 at the annual conference on small satellites at Logan, Utah, Thompson first introduced the concept of Microspace, describing technological advances that he predicted would create changes in space systems comparable to the introduction of the first microcomputers in the 1970s. A drastic decline in weight, power, and costs, brought on by constantly improving miniaturization and performance in computer technology, he said, would eventually allow new players like Orbital, for the first time, to enter an industry largely consisting of gargantuan, government-subsidized corporations that had neither the incentive nor the interest nor the vision to change their way of doing business. New companies, he suggested, could learn to compete with the Lockheeds and TRWs of the world in much the way that Apple, Inc. had entered the computer business and gone head-to-head against IBM a decade before.

The comparison to Big Blue was apt. With few exceptions the aerospace executives always required enormous capital—financial resources supplemented by major government agencies—which in turn required the assurance of enormously expensive rounds of paperwork, redundant technological systems, and several levels of oversight before they could be satisfied that most risk had been eliminated from any mission. Excruciatingly conservative designs and processes also made it nearly impossible for the industry to keep pace with the kind of rapid advances in microelectronics that spurred the efflorescent personal computing markets.

For example, traditions in aerospace established a norm requiring that most space hardware meet a military grade, a measure of testing that calculated risks of failure at infinitesimal rates. Often absurdly reductive, mil-rates could just as easily be substituted with industrial grades, which would be entirely sufficient for many pieces of hardware at a fraction of the cost. A 28-volt stack of nicad batteries, for instance, which might cost $40,000 and weigh ten kilograms under a mil-rated configuration, would cost, in some cases, only $180 and weigh four kilograms as an industrial-grade package.

For decades, the aerospace industry had allowed itself to be hog-tied by such traditions, in part because the cost of a rocket launch (between $50 million and $250 million) made any gamble with a deficient part seem ludicrous and politically dangerous. On the other hand, if launch costs were not so high, then a person might look more readily toward the example of personal computers as a reasonable model for doing business. (In the personal computer business, the industry generally followed advances in semiconductor technologies, and the rules for success could be identified, most simply, as: continual product development; a new family of products every one or two years; ever-increasing functionality and ever-decreasing costs for the same level of function; and finally, volume based on demand in mass markets.) The relevant question for aerospace was, of course: If the cost for launches could be lowered considerably, couldn't the aerospace industry also take advantage of revolutionary developments in micromechanical and microelectronic systems?

"It's a simple idea, really," Thompson said at the time. "High technology entreprenuership works. It works really well in a lot of other industries, so why not try to make it work in space?"

The usual answer, which he had once contemplated seriously as a graduate student at Harvard Business School, was that financial barriers were much too awesome for any small company. A great idea was never enough without the support of a gargantuan budget and well-cultivated political cronies. An inexpensive new satellite or an excellent business plan still had to overcome upfront cost of tens of millions of dollars just for an initial ride to orbit. To purchase a ride on a Titan rocket, designed to lift extremely heavy payloads, or even on the space shuttle, when it was in service, kept the entrepreneurial mind at bay.

"Let's say you're a scientist and you want to do global change studies," Thompson explained. "So what are you going to do? Do you go out and buy a $250 million Titan 4 rocket and an $800 million space platform that's got twelve instruments on it? I mean, you've spent a billion dollars right off the bat. It would be like, in the absence of a personal computer, you went out and bought an IBM mainframe to do your taxes at the end of the year.

"But let's say you decide to go ahead anyway and spend the billion dollars. The next thing you discover pretty quickly is that you can't carry your work home with you that night. In fact, you'll probably have to wait for another eight years to get into space—eight years! Suddenly you realize, if you want to do global change studies with this "IBM mainframe/space model," you're gonna go broke and lose so much time, it will drive you crazy.

"So you take a deep breath, stop, and reconsider the problem: 'Do I really want to put a billion dollars of eggs into that one basket? What if the rocket blows up? What if a solar panel fails to deploy?' Then you wonder if a big study is really worthwhile, and more than likely you will conclude that maybe there are smaller projects you could do—faster, cheaper, better, using little systems, smaller satellites, smaller rockets, smaller payloads—that would give you the same dollar-per-bit performance and be up and running in a couple of years with an upfront cost of $30 million instead of eight years and a billion dollars.

"At that point the analogy to IBM pretty clearly puts you in with that group of first customers who bought all those first Apple IIs. And that is how we began to think about the concept of Microspace—finding a way to provide a level of affordability and ease of use with space products that make smaller systems attractive to a much larger and more diverse customer base."

Careful reasoning convinced him and his colleagues, at least initially, to consider building their own rockets. Before even contemplating a global constellation of satellites or anything else, the Orbital Sciences Corporation would have to invent a rocket that a little company like it could afford.

In 1990, a year after his presentation of the Microspace idea in Utah, one of the industry's few grand achievements occurred not under the auspices of NASA or among its prominent corporate legions, but within an obscure team of about thirty-five young engineers from northern Virginia who had built a new kind of small winged launch vehicle. The rocket, designed to begin its liftoff not from a concrete pad or even a runway but from beneath the wing of a B-52 at forty thousand feet above the ground, was roundly dismissed as just another quirky, blue-sky project when the concept was first unveiled. But on April 5, 1990, with former NASA astronaut Gordon Fullerton piloting the B-52, the Orbital company—"a bunch of guys," according to the project's chief engineer, Antonio Elias—demonstrated the first successful expendable launch vehicle in this country in twenty years. Not since the creation of the space shuttle had any American company produced a new rocket that worked. Best of all, the rocket boasted a sticker price almost 90 percent less than the next-cheapest launcher in the country. The highway to Microspace opened wide.

And so it was that during an unprecedented low point in aerospace history, Thompson and his colleagues emerged as America's first space entrepreneurs by simultaneously inventing a surprisingly sassy young company and a surprisingly effective new rocket they called Pegasus. As equally stunning achievements, the Pegasus rocket and the little mom 'n' pop space company that built it won the National Medal of Technology that year. Orbital very quickly became the world's first publicly traded commercial space company, the subject of Harvard Business School case studies, and a sought-after supplier for the beleaguered NASA, whose new director wanted desperately to learn how his own people also could begin to do new things faster, cheaper, and better.

At least initially, the Microspace blueprint seemed to offer an antidote to the industry's malaise and a juggernaut for Orbital into contract competitions against major players. After nearly a decade of effort, the men and women who comprised Thompson's maverick ranks finally impressed NASA and its affiliates so much with their engineering expertise, visionary ambitions, and financial acumen that hundreds of millions of dollars in contracts began flowing in their direction. Pent-up demand for rocket rides to orbit were about to explode in the space shuttle's absence, and as soon as their inconspicuous little Virginia space company entered the marketplace with its own inexpensive launch vehicle—"at the price of a pickup truck," they would say, "not an eighteen-wheeler"—customers lined up. Pegasus was not a product—Orbital was not a company—that the rest of the industry could ignore for long.

By 1992, when I moved from my home in Connecticut temporarily to Reston, Virginia, to begin observing what promised to be the first boom in space commerce—"bringing the benefits of space down to Earth," as the Orbital motto went—Thompson's engineers already had a half-dozen new development projects under way: Pegasus-XL, a newer, more powerful version of the Pegasus; Taurus, a mid-sized launch vehicle; Seastar, an Earth-observation satellite for NASA; APEX, an experimental satellite contracted by the U.S. Air Force; TOS, a transfer-orbit vehicle that would propel the long-awaited Advanced Communications Technology Satellite (ACTS) into geosynchronous orbit and later drive the Mars Observer on its planetary trajectory to the red planet. Finally, there was Orbcomm, a global communications system, a completely commercial venture funded without a penny of government money, that would put into the hands of any person at any place an inexpensive mobile communicator that would connect one human being with any other human being on the planet.

The scope and ambition of Orbcomm not only made it the most interesting undertaking of all, but it also represented hopes that Orbital could beat a path into satellite heaven—the new frontier of low-Earth orbit—before any of the usual aerospace behemoths pounced upon it and claimed it as their own. And with the extravagant profits that would result, Thompson hoped that the example of his company over the next two or three decades would prove that the commercialization of space could occur without government financing. The next revolution in aerospace would happen only if, as he liked to say, "you plug the energy of private capital into space technology and then get out of the way."

Without subcontracting services for satellites, rockets, software, earth stations, or network communications systems, Orbital executives expected to create their own vertically integrated company whose subsidiaries and divisions would supply parts for every aspect of the business. As much a part of their cosmic value system as Kepler, Faraday, Newton, Heisenberg, Gödel, and Feynman, was a belief in the pure combustible energy of the Invisible Hand, a force that, combined with the algorithms of rocket science, would perform as a kind of lox generator for a new vision, giving their corporate dreams hyperbolic ignition, like hydrogen peroxide and sodium permanganate in the turbine drive of a V2, creating for them a new empire of global proportions.

Thompson jumped up and his voice lifted in excitement. He wanted to explain the scheme to me.

We had been seated at a table about ten feet away from a wall and shelves where he stashed a variety of rocket models, placed a variety of crystalline spheres and globes, and hung paintings of space probes. If you could think of his office as a scale model of the solar system and, if by rough scale, the distance from the table where we sat to the crystalline globe on a far shelf was about 22,000 miles, then the basic ideas fell into a visible arrangement right there in the room.

At 22,000 miles, he said, we were rotating in the famous Clarke orbit. That is, if the crystalline globe ten feet away represented the Earth and we were two satellites, our orbit would be considered geostationary, a placement in space imagined half a century ago by Arthur C. Clarke, the science fiction author who first conceived of three communications satellites stringing a web of electromagnetic waves that would unite the world.

We did not move; the globe did not move. *Geostationary.* In other words, just as most communication satellites today raced in geosynchronous orbit tens of thousands of miles an hour above Arthur Clarke's home in Sri Lanka—the rotating Earth appearing to stand still from space—so we were now looking at the crystalline globe, our heads spinning as the globe spun, until we were "looking" directly at one another face to face, as still as a new morning.

"Okay now," he said, "down to Earth."

The geography shifted as he moved across the room until he stood just to the right of the globe. There, only inches above the glossy surface of the Earth, he motioned to the new domain. By scale, if our heads were, moments ago, bobbing around in geostationary orbit at the table, then his little Orbcomm satellites—laced over the Earth—would circle, as he said, "right here," placing his hand four inches above the slick, round surface. In low-Earth orbit, satellites would whiz around the globe once every hundred minutes, whirring like hornets around a nest.

Miniature satellites whirling in low-Earth orbit, he predicted, would deliver the holy grail of satellite communications, what one of his colleagues called "the last big cookie," the Dick Tracy–wristwatch global communicator, the Buck Rogers, *Star Trek* handheld devices of yesterday's science fiction.

The scheme, therefore, depended not on geosynchronous orbits tens of thousands of miles away, but on this less exotic, more domestic regime—a ten-minute rocket ride from home—in a place people were calling LEO.

Instead of placing three satellites 22,300 miles above the surface of the Earth, as Arthur Clarke suggested, aerospace companies would launch twenty-five to perhaps hundreds of small satellites into low-Earth orbit, distributing them in various configurations that comprised, for all purposes, a kind of artificial constellation.

"And you might say, 'Well, that's interesting. That's different,'" he said excitedly. "And maybe it is, because no one has ever used low-Earth orbit for commercial global communications. But what becomes more interesting is when an application of simple physics shows what is gained by stationing satellites this close to the Earth."

The mention of space physics—where up was down, where falling objects may attain orbit, where mathematics cuts a trail the way a hound tracks a rabbit—was a warning to listen carefully. The little universe in David Thompson's office was about to tilt; the gain of which he spoke was worth unfathomable amounts of money.

"Common sense," he said, "tells us that if I'm off in a remote area somewhere and I can't use a terrestrial telephone system, then I have to use a satellite. Believe it or not, as much as 80 percent of the world's population has no access to regular phone service. In many places a single phone call extracts an enormous price.

"Physics tells me the power I need to make a link from my little pocket phone or wristwatch to an orbiting satellite scales something like the square of the distance between the Earth to the satellite. In other words, if my satellite is 200 miles up versus 22,300 miles up, in distance, I'm 100 times closer. But in terms of power, I'm 10 to the fourth times closer—10,000 times closer. That's an enormous benefit. That becomes a major reason to station our satellites close to the Earth.

"Of course, you might say I could partly offset the disadvantage of higher orbits by carrying a huge antenna either on the Earth or the satellite to get better gain. And for thirty years some companies have done just that. But you should see the size of their equipment—you need two backpacks and a small truck just to carry it around. So I can't carry it in my pocket, anyway, and probably I can't even take it on my back. It's hardly mobile. And the equipment would cost thousands of dollars.

"But if my satellites are flying closer to the Earth, I might not cut the power by 10,000, but I could bet on at least a fiftyfold reduction in power.

And that turns out to be a good number. At that level everything in the system shrinks."

He dipped his hand into a shelf underneath the globe and pulled out a bland little brick speckled with rows of push buttons.

The device, known affectionately around Orbital's offices as an "Orby," felt so cool and smooth to the touch that a moment passed before I realized the little bugger was only a demonstration model made of wood. In 1992 the first global communicator existed solely as a designer's toy, a primitive mock-up. Several numbers and letters on the alphanumeric pad were scrubbed off from excessive handling at presentations from Geneva to Johannesburg, but Orby still fit comfortably in my palm. It felt snug there, like the remote control for a VCR.

Let your fingers stroll across its face.

Scroll . . .

Tow . . .

Med(ical) . . .

Find . . .

Sat In View . . .

Transmit . . .

Calc(ulate) . . .

Msg Rcvd . . .

Take one in hand, and the imagination expands playfully. B-E-A-M M-E U-P, I typed.

I continued to fiddle with it while he talked.

So it was the enormous reduction in power that made all the difference. Instead of needing to generate a lot of power onboard their satellites and on their earth-based transmitters, he said, an Orby would operate fine with just a five-watt transmitter using familiar radio frequencies outside the FM band. Using such a relatively low frequency, the company could build its transmitters with existing commercial parts, mass-produced microelectronics, not unlike those found inside an ordinary portable radio.

The use of commercial electronics gave him an incredible advantage over other systems. Without having to create custom parts that would conform to an exotic stretch of bandwidth, Orbcomm's space products would rely on the world's storehouse of commercial radio parts, where were, by comparison, ubiquitous and inexpensive. They would not require years of specialized testing because they had already been tested in the marketplace. Orbital would not need to turn to a specialist to manufacture the devices because companies like Panasonic or Sony had the wherewithal to crank them out like popcorn.

"All this means," Thompson continued, "we can sell these things on a mass market at prices between $50 and $200. We have gone straight to the consumer with prices people can afford, and suddenly we're bringing the benefits of space down to Earth. In a few years you'll be able to go into your local K-mart and say, 'I want one of those red ones to carry in my backpack,' or you'll go to your local car dealer and ask for the $200 emergency messaging option for your car. In the first couple of years, we expect to break even with business in the United States alone, but by the year 2000 we project annual revenue of $1 billion on the world market."

I noticed when he stopped, as if playing back what he had just said, an expression that I thought might have reflected a moment of doubt. Or maybe it was an expression of amazement and surprise that crossed his face. Just briefly, I thought, young Mr. Thompson was about to hyperventilate on his own expectations.

He did say, after all, one *billion* dollars.

"What? You've never heard of a little mom 'n' pop space company before?" he asked.

Then, grinning blithely: "Well, we'll be the first."

A few miles away at the Sullyfield Office Park in Chantilly, Virginia, a plastic garbage can filled with ice chilled sodas outside a transparent clean room in a huge, blinding-white high bay. A handful of engineers shared a stack of pizzas around the first wooden mockup of their little satellite. The model, small enough to fit on a desktop, looked like a high school science fair project.

"This is the actual size," one of the engineers explained. "We're gonna launch a bunch of these little suckers with our Pegasus rocket. It'll hold up in this package configuration until it gets into low orbit, then it basically springs apart like an artificial Christmas tree. We can't tell anybody what it looks like, but as you can see, it's a lot like the thermal oven the delivery boy was carrying when he brought the pizzas. Everyone keeps saying the design's proprietary and top secret and all that, but once we go public, Domino's will probably try to sue us."

They called them pizzasats. Eighty pounds of carbon fiber, silicon wafer, aluminum honeycomb. Forty inches across. Six inches wide.

"Deep dish," someone said.

The team itself looked less like a group of engineers than like a college fraternity: bushy-tailed, bright-eyed, well-scrubbed sons and daughters of engineers, some of whom wore gold-plated Beaver rings—MIT alums. The lead systems engineer for avionics had just graduated from Duke. The propulsions guy was a tall skinny kid straight out of the University of Illinois. The attitude-control system rested in the hands of two young Amer-Asians, college grads of less than one year. Their conversation strung together an excited blur of acronyms. The air reeked with youthful enthusiasm.

Of the 40,000 job applications that had flooded Orbital during the past year, filed by hungry aerospace cowboys from Sunnyvale, California, to Greenbelt, Maryland, David Thompson had offered a handful of mostly inexperienced college grads these plums. In any other aerospace company, these kids would have already disappeared among gray legions of specialists to begin training. But Thompson wanted them not so much for what they knew as for what they did not know.

The engineers anticipated turning their hands toward making a new technology and creating something equally new and remarkable—a twenty-first-century space culture.

The smell of hogged-out aluminum and glue mixed with the fragrant pizzas; the sounds of whirring drills interrupted the wail of rock 'n' roll.

From the day I arrived—in the high bay, around the coffeepot, after conferences, at bulletin boards—everyone from vice presidents to engineers wondered whether the nascent experiment would flourish.

Could the Microsoft idea work in Microspace?

In conservative northern Virginia, on the outskirts of a dwindling, dismal culture of black budgets, aeronautical profiteers, and bureaucracy, it was sometimes hard to know just what to think about David Thompson and the dreams of his young engineers at the Orbital Sciences Corporation. But if thinking big would increasingly mean thinking small, the fact was, in January 1992, everyone under his roof that day imagined a moment in the not-so-distant future when bits would flow from space like crude oil.

He had given them only eighteen months to stake out their piece of the sky, and no one sounded the first note of doubt.

Bulletin Board

"You don't need a cast of thousands. You don't need gigabucks; you need megabucks. And you don't need a long, long time. You can do this in a couple of years."

—David Thompson
CEO and cofounder
Orbital Sciences Corporation
January 1992

"We could do better opening an Italian ice cream store in Georgetown. We tell each other that sometimes, but that wouldn't be as much fun and not as important. We want to do something important."

—Bruce Ferguson,
COO and cofounder
Orbital Sciences Corporation
June 1992

"Orbcomm is the one great cookie left. For the average person who has not knowingly benefited from Space, this is it. We'll have mothers calling us up saying, 'Thanks for saving my son's life.' Orbcomm will give everybody in the world an emergency service they can take anywhere—for fifty dollars! This will be a big hit. It will also bring this company its greatest financial rewards—easily a billion dollars a year."

—Bob Lovell
Space Systems Division president
June 1992

"My dream is we evolve into an organization which is as new and novel and useful to humankind as the original Orbital group was. My fear is that we don't make it. My fear is we fail in the classical sense. That is, we make so many mistakes that we lose our customers. Or we have a more subtle failure where we don't lose our customers but we grow and evolve into yet another aerospace company like Martin Marietta or Lockheed or Boeing. Even though the world does not recognize that as a failure, in my personal evaluation that's as bad as if we fold up."

—Antonio Elias
Chief technical officer
Orbital Sciences Corporation
June 1992

Orbital Contracts Awarded in March 1991

NASA, seven Pegasus launches:	$80.0 million
U.S. Air Force, small meteorological rockets:	6.5 million
Satellite Development, APEX:	9.6 million
DARPA, Pegasus rocket:	7.5 million
NASA, ocean imagery project:	43.5 million
Total:	$147.1 million

"The company's methodology is modeled after Silicon Valley's approach to product innovation: the use of relatively small teams of exceptionally bright, talented employees; pay scales slightly above industry average and long hours nurturing new ideas from concept to production in the shortest length of time. So selective is Orbital in its hiring that it will screen about 60,000 applicants by year-end to fill no more than 275 job openings. Orbital takes about thirty months to get a new idea from the drawing board into manufacturing—roughly half the aerospace industry average."

—*Aviation Week & Space Technology*
November 30, 1992

"Yeah, it's a small team. The downside is, it's easy to have single-point failures. I mean, if someone gets hit by a truck, we're screwed. The upside is, we don't need a lot of meetings."

—Dave Steffy
Satellite program manager
June 1992

"Bringing the Benefits of Space Down to Earth"

—Orbital slogan, 1992

"Vital Communications Absolutely Anyplace on Earth."

—Orbcomm slogan, 1992

"All space activity but communication satellites require government subsidies. The prospect for the commercialization of space is even more remote now than it was twenty years ago."

—Alex Roland
Professor of aerospace history
Duke University
1992

CHAPTER TWO

July 1992

Wolf Pups & Freshouts

Mike Dobbles strolled through the high bay jostling a few small silver valves, clenching under his arm a black propulsion tank that looked like a bloated kielbasa. With the free hand, he punched in a code to unlock the door to a separate bay, then stamped his feet on a sticky black mat that cleaned the soles of his shoes. When the lock clicked, he pushed his way through.

"*Pssst!*"

He glanced over his shoulder.

"*Pssst!*"

Slouched around a workbench, a smirking crew of grizzled technicians laughed and winked.

"They always do that," he said to me, then let the door crash shut.

Although he tried to be pleasant and professional by smiling and nodding to his new colleagues every morning, nothing could protect him from their challenging sneers. Any slight jitter or awkward step as he approached the first test setup, and they would hiss. A youngster's struggle against the steep pitch of his first learning curve never failed to amuse the company's craftsmen. The crew of older men—chaps with dirt under their fingernails and decades of accumulated aerospace experience—registered comic interest by mimicking the sound of Dobbles's tiny satellite thrusters.

After a few months on the job, by July '92, young Mike Dobbles could no longer hide his hand. He had slipped inside the Orbital Sciences Corporation without even a modest amount of practical knowledge. He had never taken a real engineering lab course in his life. He couldn't weld. Electrical circuit design escaped him. An oscilloscope might as well have been a periscope when he first bellied up to the bench. When he accepted the job of "specking" the propulsion system for the new Orbcomm satellites, Dobbles later confessed to me, he knew he would have to write some kind of report, but he could not fathom exactly what.

"What's a specifications document?" he asked.

Of more than 40,000 applicants for fewer than a couple of dozen new jobs at Orbital's headquarters in northern Virginia, Dobbles had managed to snag a premier spot—right out of college—designing and building the propulsion system for the world's first commercial satellite constellation. He felt grateful. Without Orbital he might still be back in the Midwest or, at best, inside some bloated aerospace bureaucracy serving an apprenticeship to an anonymous midlevel engineer who had dawdled away a decade designing a bracket or a screw for a mission that would never fly. The same sense of gratitude helped him ignore the hissing when he went to the bench each day and faced an impossible job.

Executives at Orbital had hired college graduates intending to create a sense of perpetual youth inside the corporation. The average age of Orbital's employees was down to thirty-one by July, and David Thompson had succeeded at jump-starting several small teams with a few very smart young people who, he believed, could outperform more traditional teams populated by gooey layers of middle-aged engineers. Without firmly fixed ideas about how the world works, he would say, his first recruits would be more likely to imagine creative solutions to complex design problems. They were, after all, the first generation of engineers that had grown up with personal computers, and they performed most of their work quickly and quite naturally at plastic keyboards. Given a flat organizational structure, with only a program manager to lead and all other jobs being equal, Orbital's small protected teams would not bog down in an obnoxious and inhibiting social hierarchy, where engineers constantly battled for raises and jockeyed for middle-management jobs. By allowing unparalleled freedom in their craft, it was thought that David Thompson's youngsters would quickly impose their own cult of sacrifice. Already inclined to eat, live, and breathe engineering, they would eagerly hit the brisk pace needed to take a satellite development project from design to launch in eighteen months.

For a guy like Dobbles, who had come east from a little town in Illinois and a state college in the Rust Belt, sudden immersion into the project was electric. Since springtime, he had found himself sharing lab space with a bright assortment of a dozen hip young college grads hiring in from Stanford,

Duke, Caltech, Berkeley, and MIT—exactly the kind of intellectual crowd he had longed for since childhood.

As a class, they were called "Freshouts," and though the term sounded slightly disparaging, Dobbles never took offense. In fact, he seemed delighted.

In theory, his tasks were simple. As he explained it, he would design the Orbcomm satellite's propulsion system to send a burst of gas from two tightly wrapped carbon fiber tanks ("packed at 6,000 pounds per square inch") connected by a pipeline to a pressure transducer ("for reading the pressure"), leading to a filter ("for clearing any dirt particles from the gas"), to a valve ("to control the flow"), to a tiny nozzle ("approximately the size and shape of the tip on a mechanical pencil"). Open the valve—*psst, psst psst!*—and the small satellite would leap suddenly to a higher orbit and earn a little longer life in space.

The concept was childlike; the timing begged for an expert's touch.

For all his effort designing parts and hiring vendors for the crucial subsystem, Dobbles's attention would at last focus on the production of a few precisely tuned squirts of nitrogen gas—shots his colleagues insistently called "mouse farts."

The phrase amused the technicians. When he tracked them down in the afternoons, his brow knit, his hands clutching crude sketches, asking for tips, they would snort and laugh, then leave him to ponder an absurdly belittling task: *How can I build a contraption to characterize the propulsive force of a mouse fart?*

Unfortunately, Dobbles was still a beginner, barely equal to a challenge of mousy proportions—for all purposes, an amateur in a very demanding profession.

He stood a little over six feet tall, as thin and lanky as a minor leaguer, with a heavy jaw and a narrow gray face, looking barely twenty-one. He was awkward and gawky like a boy on his first railroad ride away from home. Under trim brown bangs a thick set of dark eyebrows set a straight, uninterrupted course across his forehead, shadowing a pair of sweet-looking eyes that had become harder and more focused by the day.

As he walked around the lab table preparing for his first tests, the blast of power tools cut across the building. The high bay, an enormous concrete-and-steel structure that looked like an unfinished auditorium, amplified noise so a small hand drill screeched like a buzz saw, a staple gun yelped like a jackhammer. At times the shrill sounds made him wince.

He stood over his own handiwork, which had evolved in a few short weeks into a strangulated mass of strain gauges and rubber tubes. Their intricate windings linked a tiny silver nozzle that looked like the tip on a mechanical pencil to a circuit card and a hydrogen tank shaped approximately like a torpedo. The wires and tubes crisscrossed a supporting stage, which held the nozzle in place atop a shiny aluminum mount, a platform where the force of each gaseous mouse fart could be measured. Forebodingly, the setup actually looked a little like a scale model of a guillotine.

I watched him unpack a set of handwritten notes on grid paper after twice circling the lab table. He firmed up connections on each of the wires, fumbled with his sensors, then puzzled over where to attach a set of strain gauges.

His face twisted.

"How do you measure . . ."

He pressed the button in his hand—*pssst! pssst!*—the gas tank fired two quick shots through the tubes and the nozzle popped off its mount.

"Anybody could understand this," he mumbled, hooking wires to the tip of the nozzle again. "It should be as easy as letting air out of a balloon."

He pulled a stool up to the bench and sat down.

"Wait," he said, "wait a second. Where is that damn thing!"

Rummaging around, he finally found the push-button control resting in his lap.

Pssst! Pssst!

Dobbles's fingers flitted over settings on a voltmeter and power amplifier.

"It's probably good that people can't read what's on everybody else's minds around here," he said.

He peered into his handcrafted cage of snaking wires and rubber connections. Staring into the center, he again eyed the miniature nozzle. His brow plunged like a cosmonaut's at ten Gs and he reached back to set the dials on an oscilloscope that would measure the burst to the nearest microsecond. Raising a fingertip to the gas control, he rapidly tapped the button.

Spastic jolts jerked the tiny propulsion nozzle.

The spacecraft went tumbling through his thoughts.

When he looked up, brow wet, a young engineer at the next table over was laughing. Strands of smoke curled out of Dobbles's microprocessor.

"Nice job!" smirked the neighbor, standing with his own test setup before him.

"Yeah," said young Dobbles, not amused. "A regular barbecue."

Steve Gurney zipped back and forth between lanes on the highway, gunning his silver Saab to dodge slow commuter vans trundling toward the exit lane. He did not like the look of the line at the tollgate.

Driving from his apartment in Georgetown to his job in northern Virginia took about forty minutes, a little longer in this kind of traffic. The view beyond the herding masses of automobiles had changed from Georgetown's clutter of urban boutiques and traffic squalls, which he preferred to navigate on bicycle, to a broad complex of flooded beltways and toll roads linking Washington to its affluent northern suburbs.

His mind raced and passing views escaped his attention—road crews and construction teams, staunch office towers whose glassy profiles and nouveaux designs spoke boomingly of the area's wealth and ambition. Fairfax County had lost 40 percent of its trees over the last two decades and increasingly supplanted its sylvan landscapes with strip malls, apartment complexes, and office parks. As a native of the region, Gurney hardly noticed except to feel surrounded by options. Life fairly buzzed with opportunity if a person made the right choices. Gurney had, naturally. He was a software engineer.

Commuting east into Maryland, to Goddard or Greenbelt, as he had in previous jobs, might have unsettled his life even more than the drive to Sullyfield. The only other consolation would be if the daily trip eventually forced him to buy a motorcycle. More speed, more fun, a high-velocity rush would not only clip his road time by twenty minutes, but a 90-mile-an-hour burst up the toll road would buffet him awake in the mornings—just what he needed after long evenings mashing bit-packets through a satellite's first lot of microchips.

He pulled into Sullyfield a few minutes after nine, nosed into the parking lot outside the Orbital building, which was already crammed with cars, and made his way toward his own windowless quarters on the first floor. His partner, Rob Denton, had probably arrived in the lab at seven-thirty as usual and gone straight to work on the flight computer, the electronic traffic cop of their smaller cheaper, faster communications satellite.

Denton, "Mr. Sunshine," was already upstairs twiddling with his college ring, simultaneously laughing and scrutinizing the first motherboard while also chatting it up with the project manager, who usually stepped out of bed at five-thirty and arrived as early as eight.

Although his partner's disgustingly bushy-tailed attitude was no match for Gurney's own bleak and ironic take on reality, Denton did make for an interesting benchmate. The consistent peak of energy he hit every day by midafternoon kept their work always on a fresh pace. Gurney would never admit it to the rest of the team, but Denton did have an addictive effect. In fact, the whole scene at Orbital was addictive—and liberating. They reveled in their work as engineering inconoclasts, bred to disdain formal authority and bigness and slowness and costliness—for all too long the watchwords of the old space age.

In his office he had tacked up a poster that said *Kein Angst!* "No Fear!" in German. His German girlfriend—*former* girlfriend—had taught him the phrase, which now registered to him as applying as much to his work as to his social life. Nearing his twenty-sixth birthday, Gurney suspected that in terms of pure survival adulthood could easily be summarized by five essentials—basketball, good work, a steady girlfriend, a big city, and a steaming pot of coffee. He could see how a person might live contentedly with those for at least a few years, until his knees wore out or—God forbid—he married. He expected that, for brief stretches, hard work alone might even be enough to sustain him.

Except for the poster, his office was a dreary pit on the first floor—no windows, white walls, desk, computer—unlike some of the nicer digs upstairs. It hardly mattered, though, not that he was oblivious, but nobody spent much time alone in their offices. The excitement occurred upstairs in the second-floor lab, where his team engaged in communal work that kept everyone busy until after dark. Continuous action, unlike any he had ever seen, made this job especially attractive.

The pace made the work exhilarating, in fact. Their daily twelve-hour struggles to push bits through the satellite's first lot of idiotic microchips passed with pleasure. The project was like a cowboy ride into the next millennium, like a hack into the exotic Infobahn, not a dreaded expulsion into some link-a-bit base of bureaucratic smegma where project managers kept schedules posted daily and everyone was boobishly conscientious. Gurney felt like he had entered a charmed circle, addicted to risk and challenge and all-nighters, when he stepped in to work every day. As cool as Silicon Graphics, as spicy as Apple ever was, this gig was, as he liked to say, *very sweet.*

For a twenty-five-year-old like Gurney, who had graduated from Washington University and gone directly into aerospace because it seemed "cool," the Orbital job offered a chance to work in a fast, over-the-top environment that he had expected from a career in the first place. At least for him, after suffering previous jobs subsumed in bureaucratic waste, pork-barrel politics, and union regs, his first risky venture in commercial space introduced him to a new cult of engineering where the lifestyle was as compelling as the actual work.

Every weekday morning beginning around seven-thirty, whole groups of young engineers drifted in past front-desk security with beaten leather briefcases, nappy backpacks bulging with running gear, and bagged lunches. Red plastic photo ID cards dangled from their lapels. Stocked with 486s, Suns, and galleries of CAD supplies, their offices spilled over into the noisy corridors of the second floor of the Sullyfield building like a block party swimming in home brew.

With so few people on each satellite team, formal meetings were few. The team's solitary mechanical engineer could walk across the hall and discuss a design change with two attitude-control engineers, and then they could bop down to the lab to relay the news to the lone battery guy. The traditonal need for multiple layers of documentation and long weeks of analysis might vanish in a twenty-minute conversation. With a small satellite, everyone understood the overall requirements, making peer review relatively simple and paperwork nearly nonexistent.

So by summer the second-floor hallways had become the organizing spot for cabals and technical collaborations, heavy with rumor and gossip and gales of conversation about engineering with semi-mystical overtones. The company's multiple satellite projects, lengthening manifest of rocket launches, and bottomless cache of soft drinks made the workday a celebration of sorts. Freshouts, in particular, labored eagerly long into the night.

They came to work in double-pleated linen pants and leather sandals, mock turtlenecks and rib-stitched cardigans. With their wet tufts, buzz-chop sideburns, and punky dispositions, they carried the effortless style of urban sophisticates, the caustic manners of magna-cum-laude graduates with James Dean attitudes.

They had a thing for fine cars. The young software engineer from Boulder, Colorado, spun out of the parking lot after work, overtaxing his red Miata like a teenager out to carouse. The latest hire from California toured in an antique Mercedes. The talented RF engineer—the one man who understood the black magic of radio frequencies—had his eyes on a Jag.

All in all, they demonstrated behaviors that banished them to one of the two far slopes of a Gaussian tub, boasting bold quirks of fashion and personality, odd twists of intellect or physical bearing. Top graduates of the country's most prestigious aerospace engineering schools entered the company with vocabularies rich in the lexicon of orbital mechanics and digital signal processing. Exotic concepts rolled off their tongues as fluidly as the tastiest licks of John Coltrane. Their lab conversations were as esoteric as the midnight dissonance of Monk-like chords. They spoke faster than most people, thought faster, and in their company one always had the sense that they were stealthier and more conspiratorial than others their age.

Grace Chang, the attitude-control engineer, an aggressive Amer-Asian from Stanford who had served as captain of her university's fencing team, made her way through the high bay practically dancing, singing pop tunes at high decibels. Rob Denton, the young gentleman from Duke designing the satellite's power system, had just installed sixteen phone lines into his townhouse apartment to bring up an online game service, a kind of

Dungeons and Dragons tactical challenge that he expected would eventually score him a small fortune. Morgan Jones, the bohemian-looking tennis player from Boulder, who would build the satellite operating system, was scheduled to fly a test on the next space shuttle mission to investigate the nature of solar flares.

On the whole they were noisy, quick-witted, athletic, nimble, shrewd, nervous, cocky, and brash. They also displayed a few other characteristics that set them apart, traits that had finally made them clear standouts among tens of thousands of engineers who had applied for jobs at Orbital that year: an unusual quality of intellect and imagination; an articulate, sometimes ironic opinion of what it meant to "be" an aerospace engineer at the turn of the millennium; a sassy, flaring kind of temperament that could easily be confused with egotism; and a heightened sense of personal destiny.

It was precisely the kind of place Steve Gurney had always wanted to work.

Brushing his hand through his black hair, Gurney unpacked a couple of manuals about 68302 processors, then placed a call to a company that could help him line up an Internet connection to the office. He stroked his hair again, left it tufted, yawned, and rubbed his eyes.

Not exactly a career yet.

He wore a ho-daddy haircut, black rat-stabber boots, and blue jeans. Thin and lithe, he usually met people in the hallways with a cynical sneer or an ironic turn, a cheeky urbanity that seemed to fit the company's *wolf puppy* image. That was what Dave Thompson had called them, anyway, at the last all-hands meeting—*wolf puppies*, he said "to differentiate from the other dogs." The image fit—at least it was better than *Freshout*, meaning "fresh out" of college, which Gurney definitely was not.

Unlike most of the true Freshouts, who came directly to Orbital either from graduate school or a first job, Gurney had moved to Orbital from the musty dens of traditional aerospace, straight from labs so Retro that he just assumed they had not yet emerged from the glory days of NASA. Aerospace had often seemed to him like a Neanderthal. Most places where he had worked were from a distant era, where decrepit buildings filled with tired old men and dusty mainframes supported a horde of engineers who sucked a livelihood off a fragile mass of federal subsidies, pork-barrel politics, and dwindling defense contracts.

But not here, not in the high bays where the Pegasus rocket was created or in offices where the National Medal of Technology had been replicated and handed out as paperweights for a small gang of young managers.

"Orbital's the opposite of my former jobs," Gurney told his friends in D.C. "The people are especially cool."

He slipped upstairs and ambled around the hallways, peeking into offices to see who was already at work, who had wandered in with a cup of latté from the new Starbucks. His black rat-stabbers scuffed the carpet.

Annette Mirantes, the other software engineer, had flipped her office light on, but she was gone—probably mashing bits in the lab.

The hallways looked like a bright maze in the mornings, as they turned and connected, dead-ended at a kitchen, a lab, a dark room full of design tables, another stocked with copy machines, a door that opened up directly 20 feet above the floor of the high bay.

The Sullyfield offices boasted a more refined quality than one might have expected of most start-up ventures. Large models of the Pegasus rocket stood in the lobby, fine teal-and-salmon-colored pastel carpets ran the hallways, and even an arboretum thrived under pink fluorescent lights upstairs. The workplace had all the ambience of a sushi bar.

When the first wave of satellite engineers came to work there earlier that spring, they found themselves escorted into an engineering paradise with cleanly scented rooms and empty desks and bright windows, offices stocked with new computers and unconstructed bookshelves set sideways on the floor. The company provided them with bubble-wrapped, top-quality test electronics and software packages, recently purchased to meet the expectations of the postmodern engineer.

Gradually, though, by midsummer, Gurney realized that most of their offices finally began to look normal, barren, or disheveled, unadorned by anything remotely decorative, occasionally squalid, littered with crumpled papers and worn college textbooks. In just a few months, the wolf pups had made themselves completely at home.

He stopped at the end of one hallway just around the corner from the lab and opened a door that overlooked the gigantic high bay. Mammoth thermal cycling machines, quaking torture devices for satellites, clean rooms with sewing machines and scattered pieces of power equipment—the vaunted abattoir for electronics—crackled or thundered to life beneath him.

Directly below he saw the yellow outline of a body, as if an engineer had recently bolted from the lab, leaped from the threshold, and plunged to his death on the cold gray floor of the bay. From above it looked a little like one of those Matt Groening figures from *Life in Hell.* Kind of amusing.

Down another hall he reached the kitchen and fished through the refrigerator where the company stocked free soft drinks. He eyed a stack of Jolt! colas, but then closed the door and reached for the coffeepot instead. He poured himself a cup. New appliances, new carpet, great lab space, free drinks, the best equipment, smart people—what could be better?

Cappuccino would be better. An espresso machine. Real cream.

Actually he did have one complaint.

When he first arrived for work a few months ago, Gurney had filled the only software job on the team. The first week he went about the work, as any skilled Microserf might, by drawing bubble diagrams, a visual, systematic design of the satellite's communications and message-handling functions. By tagging a bubble diagram to each piece of hardware, Gurney expected eventually to produce a visual image of the total satellite system. That way he could imagine which chunks of software needed to be written, how the message-passing functions would occur and what kind of new hardware he needed to purchase to make the scheme work. Serial ports, peripherals, the amount of RAM he needed, the amount of data storage space—everything would fall from the bubbles.

But when project manager Dave Steffy saw the beginnings of Gurney's bubble work, he called an end to it immediately. Waste of time, the boss said. We don't do bubble diagrams here. He ordered Gurney to plunge in and write the simplest drivers first, then help install the satellite operating system. Once that was done, he was told he could write the code for Rob Denton's flight computer, as it was being built. In the meantime his boss suggested that he find an off-the-shelf graphics program to run the satellites' ground control system.

Gurney was outraged. As a software engineer, he knew a design-on-the-fly plan posed unnecessary risks. He felt less confident knowing the project manager's training and experience came exclusively from aerospace, that his specialty was electrical engineering, and that he once even admitted that his own knowledge of software design had ended sometime after taking a college introductory course to assembly language. More than anything else, the program manager's age—with a range of seven years between them—put Gurney at a most serious disadvantage. The man understood software only as it had existed in the prehistory of C.

But any efforts to remind Steffy that the team needed more software engineers were met with statements like "Well, gee, Gurney, under pressure one software guy can write an amazing amount of compact code." What could he do with a person who admitted that his idea of the typical software engineer was "a cross between the Maharishi Yogi and Tommy Lasorda in his pre–Slim Fast days"?

Still, Gurney was not one to worry. Complain, yes, maybe. But worry? Not him. Neurotic anxiety was not one of his failings.

Stepping around the last corner at the end of the hallway, he entered the bright lab. Denton's laughter broke across the room. A few other members of the team already had their boards running.

"Hey, Gurney!" Annette shouted. "I'm pretending to be the flight computer talking to the BCR. What are you pretending to be today?"

He groaned, sat down next to her at his terminal in the far corner of the room, and typed in his password, the name of his former girlfriend: B_R_I_T_T.

"Gurney!" said Eric Copeland, the team's RF guy across the bench. "Have you tried a little kerosene and dropping a match to it yet?"

"Yeah, Gurney," Denton said. "Why do we need an operating system, anyway?"

Software. No one understood. Sure it was easy to write—but impossible to perfect.

Tapping a few brisk keystrokes and sipping a taste of hot java, he saw the monitor of his new Gateway scramble to life and break out into a dozen colored windows of code, graphics, and graphs.

Gurney reached for his mouse.

"Screw off," he said.

It was the start of another fine day.

CHAPTER THREE

September 1992

The First Slip

By September the project manager seemed perfectly content with chaos. He was still living with vestiges of the old aerospace culture, particularly among vendors and contractors struggling to survive under a changing financial climate in the federal government. He battled with companies steeped in the history, relationships, and old ideas that had rooted the industry in long lead times, redundant systems, and excessive paperwork. The transition was not always pretty.

Not quite like his charges, Dave Steffy personified the kind of engineers the company had hired five or six years before, when Orbital's first small team of thirty-five started building the Pegasus rocket. Educated at MIT and trained at Hughes, he was as happy to spend his weekends adjusting his car's carburetor as cleaning intake valves on a rocket.

Like most engineers among David Thompson's first round of hires in the mid-1980s, Steffy had at least ten years of experience to his credit. Before coming to Orbital, he had endured enough inertia from the inside of other aerospace companies to have contemplated professional death by boredom. Like many of his older colleagues, Steffy had spent a portion of his youth on what he called "idealistic" missions, like the Magellan and Galileo deep space probes, and he also clocked the requisite time with small ancillary pieces of a com sat and military space missions.

Although he didn't mind the Hughes corporation, in and of itself, he had grown increasingly frustrated because nothing he built ever flew. This is how he explained it when we first met, in the spring of '92:

"I got to the point where, as a long-term space cadet from the time I was six, I was really thinking about going into a different industry because of the endless product cycles. L.A. grew tiresome and, since I was eager to build something with my own hands, when I learned about Orbital and heard the company was looking for a few creative engineers to build an experimental launch vehicle, well, I jumped at it.

"Very simply my goal was to work a two-year program. And when we built Pegasus, which took two years and five months, that was pretty close. It made a good fit for me. It was a great project, it was hands-on, and it was fun. We started out with no customers . . . a totally internal program with a commercial goal—I mean, nobody does it like this in the aerospace business! And guess what? It worked."

In a way, Steffy might have been an odd choice to lead the Orbcomm satellite project. Everyone on the Pegasus team knew him as a self-described "techie" who often joked that "the word *manager* means you're the one who takes the shift nobody else wants, then all you do is raise your sword and yell 'Charge!'"

Suddenly, at age thirty-two, he found himself with the title Program Manager on his office door, lacking any real management experience, while starting the company's most ambitious project with a list of charge numbers for hiring engineers and signing vendor contracts.

One of his friends from Pegasus, another young engineer (also an MIT grad, who came from INMARSAT) named David Schoen, took charge of the overall development for Orbcomm's global communications system in a separate branch of Orbital called Orbcomm Inc. The collegial, one-for-all spirit that characterized the Pegasus program quickly turned the former colleagues into intercorporate adversaries, as Schoen became the customer for Steffy's satellites. From the moment the project started, Steffy found himself taking orders from his onetime rocket-team colleague, who set rigid specifications on power and communications requirements for the satellite. Wanting to avoid a formal contractual relationship, Steffy could do little more than promise Orbcomm Inc. that he would deliver the first two satellites on time.

It was an awkward and unfamiliar role, but he tried to reassure himself that no amount of management training could have prepared him well enough.

"I took the job, but I told my bosses that I didn't want to spend more than two years working these first two satellites," he told me at the beginning of the summer. "I've got a short attention span. I hate meetings. And I detest design reviews where everybody sits around and looks at viewgraphs until they go numb. The truth is, I'm basically an engineer at heart. So as long as we launch by next summer, everything will be fine."

The odd idea of constructing a satellite without designing it first created some unusual quandaries. Some constraints were strangely arbitrary. Under the original plan, for example, the constellation required thirty-six satellites, but the power requirements to handle all the message-passing functions forced each satellite to be so big that it dictated thirty-six Pegasus rockets to launch them all. Seeing his start-up costs suddenly escalating to $500 million rather than $100 million, as projected, David Thompson decided on the spot one day that the satellites had to be small enough to launch eight at a time from the nose of a Pegasus. There were no systems engineers or operations research people around to spend months verifying the wisdom of his choice, as would be the case in a traditional project, so with a split-second decision, the physical dimensions of the Orbcomm satellites were viritually assured.

The most vexing problem, however, was not an indiscriminate design challenge but a sudden business decision to increase the number of messages each satellite could handle on orbit. In 1991, when the satellite was first designed, Motorola's 68302 microprocessor offered the best, fastest, and least expensive way to run data from the ground through the spacecraft. With six of them aboard each satellite, the 68302s should have proved as good as the original Apple computer, which was exactly how David Thompson described the scheme: every satellite "an Apple in the sky." In essence, a personal satellite.

But in 1992, as parts arrived and engineers designed the actual hardware, the specs changed dramatically. People planning the business at Orb Inc. decided they needed more message traffic to meet their financial goals and passed along a new set of specifications to the satellite team. Consequently, the power demands on the spacecraft increased, and efficiency standards placed new restrictions on the 68302s, making them slower to lessen their draw on overall satellite power. That little change put some engineers like Gurney under the gun to make up for the losses with innovations in software.

The satellite itself also changed in its "pizzasat" configuration with fixed solar panels to what became known as the "Mickey Mouse" design, a satellite with deployable solar panels and an eight-foot antenna that sprang from a slender internal trough. The unexpectedly difficult configuration change occurred well after the satellite team had begun building up the first sets of motherboards and sent everyone scrambling to make up for new weight limits, to solve difficult space constraints, and to deal with much higher accuracy requirements for pointing the solar arrays at the sun and for pointing the antenna at the Earth.

So instead of building an elementary, passive design for the Orbcomm satellites, as originally intended, late in the springtime Steffy's engineers were busy redesigning a set of large round panels on which to apply hundreds of solar cells and creating new mechanical devices that would spring the panels from either side of the satellite like two great round black ears (hence the satellite's new resemblance to Mickey Mouse) and pop an eight-foot antenna out of a three-foot trough.

The change occurred so swiftly that Steffy's team was left with significant ongoing problems. Because Orbcomm Inc. demanded more power out of the satellites to transmit and receive terrestrial signals, the communications gear inside the spacecraft expanded physically. As with any tightly prescribed engineering project, modifications cascaded through the total system, squeezing boxes together, claiming free space inside the spacecraft, raising costs, adding weight, burdening everyone from hardware designers like Mike Dobbles to software engineers like Steve Gurney.

For the engineer in charge of generating and regulating power, Rob Denton, the challenge was almost too much to contemplate. Power on a satellite is King, the fundamental limiting factor. After all, for a commercial satellite, the function of the spacecraft is to turn sunlight into revenue. The more sunlight you can turn into electrical power, the more electrical power you can use to send and receive information. The efficiencies of doing that, saving weight, and maximizing revenues tell you how close you will come to abridging the fundamental laws of physics. It was a like a game, in that sense, to see just how close you could come to that unyielding wall set by nature.

Steffy referred to it sometimes as a balloon game—"You squeeze your little piece of the balloon, and a bulge pops out in someone else's face."

But the "game" had even more challenging facets than the laws of physics. After having agreed to an eighteen-month schedule to launch, Steffy found that the new design pressured his team in ways no one might have ever imagined, and he began to fight the new specifications almost daily. A long series of meetings, angry memos, and negotiations resulted only in more conflict, which came crashing in on his life like a season of thunderstorms.

Steffy had promised himself before taking the job that he would never bother his crew with the nasty details of management infighting. He, alone among them, would negotiate whatever political games needed to be played. The bluster of gruff office politics, which he referred to merely as "intramurals with the customer," should never disturb the engineers, he had decided, even when suspicions that the schedule would slip beyond David Thompson's ambitious expectations set the stage for the first gamely round of backbiting and sometimes ugly infighting.

One morning in late September, looking nervous and uncertain, Mike Dobbles poked his head into the project manager's office. Steffy sat staring out the window.

Contrails of two jets left crosses in the sky. The manager's white board looked like a weather map after a long series of technical discussions the previous day. Arrows, zigzagging notations, and globs of complex equations stacked in distinct rows clustered around sets of algorithms for one or another subsystem on the satellite. In the center appeared a large bell curve that he used to explain his theory of launch schedules.

The morning had not been particularly rife with good news. Steffy opened an e-mail up on his monitor saying the Israeli company making the satellites' computer chips was on the verge of bankruptcy. The people building the attitude-control electronics had decided to stall Orbital's contract so they could do some work for more lucrative government jobs. One of the software engineers had walked in before Steffy's ten o'clock appointment, shut the door, and complained that he had been hired to do systems engineering, not to write software, and he wanted an immediate explanation.

Through it all Steffy managed to shrug and joke. But now, just before lunchtime, with the day still stretching out before him, he showed signs of weariness.

"Everybody's got problems," he said as Dobbles sifted through a file of papers for his presentation. "It's no big deal. Every project starts like this. We just have a few more growing pains at this stage because some people are so inexperienced. Not everybody's calibrated yet."

He sounded like he was talking to himself, although he might have been trying to set Dobbles at ease. Or maybe he actually anticipated more trouble as soon as the young man appeared at the door. In any case, even though Steffy was privately preparing to screw up the pressure on his whole team, he still made an effort to be gracious.

"So how's the world of gas?" Steffy asked, brightly.

Dobbles mumbled something about the previous afternoon's propulsion tests.

Steffy reached back to click the e-mail off his screen. "Everybody's been giving me gas today, including *my* boss. Might as well stay on subject."

He looked like somebody's science teacher. Straight, neatly clipped brown hair brushed across his forehead and touched his ears. His eyes were steady and bright. To his young engineers Steffy probably seemed like a big brother or, at the very least, the friendly neighbor next door. They saw him most often in this clean, sunny office on the second floor, working from sunrise to dusk in a white shirt and necktie, door open, back turned, hammering out rapid successions of e-mails and technical reports. His industriousness set the tone for a small team that still seemed eager to please.

His day actually had begun optimistically at corporate headquarters with the delivery of a financial report stamped "CONFIDENTIAL." It was a short but significant document that quickly made the rounds among the executive staff. Business was particularly good, the report explained, as the first half of the year ended with peak revenues and an immense backlog of orders. J. R. Thompson, a leading figure at NASA who had helped bring the space shuttle back to launch status after the *Challenger* explosion, had joined the executive staff, and the hunt for NASA funding had already begun to improve under his influence, ladling gravy over an already impressive mound of contracts with the air force. The company was far ahead of its projected revenues for the third quarter.

The good news made Steffy especially amiable and talkative as he met at forty-five-minute intervals with members of his small team that morning. Once a conversation began, he rarely paused except when an airplane came into view on its descent into Dulles airport. Then he would arch his back and stop, momentarily, to watch the aircraft as it disappeared over the horizon. "I love stuff that flies," he would say, and then return to their progress reports.The move had a way of taking his engineers by surprise just before Steffy cornered them like a professor during oral exams.

Most of the time only one person sat with him at the long conference table, but occasionally the room was as busy as a train station before a holiday.

The deadpan look on Dobbles's face hadn't changed. He jangled a few silver pieces in his hand, then tossed them onto the table.

"Here are our valves. But first you should know I've got composite for the tanks," he said.

"Yeah?" Steffy said, perking up. That would save weight

"It's just like a fireman uses," he said. "So now when we put the system together, you'll see the valve is lighter, the tubing's lighter, the pressure transducer is lighter, and the tanks are lighter—much lighter."

"Well, I hope you can show me that everything works," Steffy said. He gestured to a mess of stapled pages held tightly in the youngster's grasp. "And make sure your numbers are accurate. We don't have time for a doctoral thesis."

Dobbles seemed taken aback, but quickly recovered as he continued: "I've done the best I can do to understand the problem, and looking at the baseline—" he started to stammer, but again caught himself—"I know we don't want it to fall into the ocean for lack of propellant. So I thought, at first, we could add this 25 percent margin—"

"I won't buy that," Steffy said.

The Freshout smiled instantly.

"Just kidding."

Steffy looked surprised.

"I actually used a two percent margin. What I did was I went back and did the analysis last week looking at the insertion point and took it line by line using two percent—and I think I've solved the problems with the valves."

Steffy's face brightened. He knew all about the ongoing problems with Dobbles's test setup downstairs.

The truth was, Dobbles said, he had discovered that his problems were due not so much to the kludgy test setups as to the company that Orbital had hired to build the parts. Evidence from his tests indicated that the orifices on their propulsion valves were slightly distorted, misshaped just enough to impinge the free flow of gas. Without asking anyone for advice, Dobbles had gone directly to the source and made a subtle but effective threat.

The vendor with whom Orbital was about to sign a contract for the valves suddenly found himself in a confrontation with a fairly cagey Freshout. Apparently, sometime in the past week, Dobbles had gotten the man on the phone and, sounding older and more authoritative than he might have otherwise, applied heat.

"I told him he had better improve his work or engage in competition with alternative sources for our business," Dobbles said proudly. "He immediately agreed to look at the valves again, and then I told him that I intended to set up a design review so I could check his work in real time. He agreed to that, too, so I should be flying up to New York in the next couple of weeks. "

Steffy listened intently. So far, so good. "Okay, and what about the thruster?" he asked.

Dobbles snapped his fingers and reached for another paper. "I've been shopping around to get us a bargain price. Luckily, I found out that one of the defense contractors had some spares like we want just lying around on a shelf. I called, and we're negotiating for a price. I think they'll sell them for half the cost we had budgeted."

"Okay! Bargain-basement deals!" Steffy relaxed into his chair. "What else you got?"

Grinning, Dobbles raised his arms in mock celebration.

"The real news is, I also called up one of the companies that made thrusters for the Defense Space Communications Satellite—DSCS III—and they agreed to my absolute minimum price."

Steffy leaned forward. The Freshout had his full attention.

Dobbles picked up the silver pieces off the table and placed a couple of the tiny valves in Steffy's hand.

"Hmmm . . . interesting . . ." Steffy held the pieces up to the light so he could look through the holes. *The thickness of a camel's hair.*

"Most of it's just been luck," Dobbles said modestly. He carefully pushed his papers at Steffy, stopping to flip to a specific page, and called his attention to a set of numbers.

Steffy ran his fingers over the page silently while Dobbles talked. Steffy nodded, sniffed, and then pushed the packet back across the table.

"Nice work," he said. Then he stood up from his chair and stretched over to the wall to adjust the thermostat—"about two dB," as he put it.

Dobbles cleared his throat.

"I do have one other item of business," he said.

The expression on the Freshout's face had turned serious again. Steffy sat down.

"It's the attitude-control system," Dobbles said.

Steffy said firmly, "It's not your problem."

"But it's an extremely *big* problem."

Steffy looked at his watch and seemed agitated, but Dobbles pressed ahead, presumably thinking that he had earned the right to be heard.

In the past few weeks, Dobbles had been increasingly insistent that, according to his analysis, the satellite lacked sufficient control authority to prevent even the slightest thruster pulse from sending the spacecraft on a disastrous tumble. The combined effect of his thrusting maneuvers with other disturbances on orbit would defeat the spacecraft's ability to send and receive messages.

If he had to be responsible for on-orbit maintenance—for raising and lowering the satellite to correct its orbit—Dobbles wanted someone else to give more thought to the attitude-control problem. The one young engineer who had responsibility for the attitude-control system, Dobbles believed, really had no idea of the terrible implications of their essentially flawed design.

"This is a big problem," he repeated. "We need to make some major changes—add momentum wheels and a gravity gradient or something. Either that, or the satellites are going to tumble. Eventually we'll lose complete control of our spacecraft."

Steffy listened, but just barely. It wasn't as if he thought the attitude-control system was unimportant. The engineer assigned to the task, a recent master's graduate from MIT, had conducted hundreds of hours of simulations. Data was flooding in. At the same time it was not appropriate for

Dobbles to indulge his own anxieties by shifting attention to someone else's work. Besides, they couldn't afford to buy a momentum wheel.

Just as Dobbles began to stress his argument more forcefully, two other engineers appeared at the door with their reports.

"There's no need for you to worry, that's my job," Steffy said. He motioned the others in and waved Dobbles out. "Give it a break, Mike," he said. "You've done a good job today."

The next pieces of satellite hardware hit the desk for inspection, and the project manager continued without pause until well after midday. He might as well have been rolling up his sleeves, preparing to disgorge parts of his own piston engine in the garage at home—another exercise that would have given him pleasure.

Meanwhile, downstairs in a walled-off cubicle outside the high bay, Dobbles was brooding. He wasn't the smartest or fastest or most experienced engineer on the team. But he couldn't help believing that everyone else was moving a bit too fast—maybe they were all a little too smart. In fact, some people at Orbital might not be quite the prodigies they were made out to be.

The Freshout was fully prepared to play hardball if he had to. Already he felt that the future of the project was at stake, depending on his own savvy intervention.

"I could kill them!"

Grace Chang charged into Steffy's office, bouncing her attitude-control electronics box around her shoulders as if it were a shot put.

"They built this for an armored tank, not a satellite! My God, it's twice as heavy the one we wanted!"

"Fresh out" of Stanford University graduate school, Grace could deliver a ferocious report if she was riled. Black hair flying, fists shaking angrily, she stomped around in a stormy fit, something the project manager had seen before.

"Grace is great," Steffy said to the next engineer. "A vendor steps out of line, she rips their ears off."

But Tony Robinson did not want to hear about Grace Chang's problems. He held something precious in his hands, which Steffy recognized as the latest version of an antenna model—a homely contraption made of three stretches of white styrofoam, connected with thin splices from a yellow Stanley tape measure.

Steffy sometimes missed the nuance of his mechanical engineer's presentations. As a young Englishman, Robinson talked with a hint more Cockney brogue than upper-crust lilt, and since he also spoke rather quickly, he often sounded as if he were trying to sell something rather than demonstrate a particularly smart idea behind his rough-cut models, which he cranked out steadily and with apparent ease.

Robinson was keen on physical demonstrations. He believed they tended to support his case and helped him demonstrate complicated analytical problems quickly.

He set the object gingerly on Steffy's conference table, folding it from one long eight-foot piece into three segments, and then packed it together like a collapsible fishing rod. He had boasted that the new antenna design would result in an original, potentially patentable object. And here it was: a large device that folded without the use of ordinary bolts and hinges and would stow inside the satellite's exceptionally short and narrow trough.

"Look here, I sprang it on my wife last night in the living room," Robinson said excitedly.

Connected by six pieces of Stanley tape, which served as flexible, bendable "hinges," the antenna boom broke down into a sandwiched set. When Robinson released it, the pieces sprang apart, unfurling neatly across Steffy's desk. The project manager laughed out loud.

"It's ingenious!" Steffy said.

"No sir, it's prrrr-fect!"

Hardware continued to hit the table with each meeting into the afternoon: the glistening jewelry of electronic circuits; small towers of microprocessors rising out of green, cracker-thin circuit boards; a set of black batteries the size and shape of small, hand-sized water balloons; a new GPS receiver wrapped in silver foil; a custom-designed frictionless hinge. Each piece underwent examination at his conference table like a biological specimen in a lab. As his engineers argued physics and used strict calculations and crude drawings to explicate their ideas, Steffy added to his list of "THINGS TO DO," scrawled haphazardly down one edge of his whiteboard.

When Rob Denton came into the room, the project manager's mood actually brightened considerably. Always sunny, politically astute, Denton brought an optimistic tenor to their meetings. He also produced neatly typed records of his work and peppered his weekly status reports with exclamation marks, highlighting what appeared to be genuine progress in the lab.

Steffy grabbed an apple and a sandwich out of a paper bag he brought from home as Denton sat down. It was already three o'clock, and he hadn't eaten lunch yet.

"I know you were saying we'd have to integrate electronics in the first Orbcomm model this week," Denton said. "But it's going to be more like Christmas Day."

"That's a three-month hit!" Steffy exclaimed.

"Yeah, well, I only picked Christmas so we could all eat turkey," Denton said, smiling. "It's really going to be worse than that."

The Freshout eased his report across the table.

"So if the launch is scheduled for next July," Steffy said, examining the paper, "this puts us into December '93."

Looking not only at a schedule hit but at an enormous increase in the cost of their satellites, Steffy knew immediately that his own bosses in the Space Systems Division would not be pleased.

He thought a moment, brow furrowed, and said: "Reality is, no one knows what the schedule is a year from now. That's just a fact of life. But the other fact of life is upper management doesn't like to hear about something like this."

"Exactly," Denton said.

Steffy looked puzzled for second. The Freshout was unflappable—and this was such dismal news!

Or maybe he did understand, after all. Perhaps here was one Freshout who would not require so much *calibration.*

"You know, bridge builders and highway builders have the same problem in spades," Steffy continued. "People just tend to read more about schedule slips in aerospace than anyplace else. I'm always surprised because you never read in *The Washington Post* that there's been a terrible cost overrun on Highway 28 and a massive schedule hit. You just take it for granted that most of those guys are going to be out there leaning on their trucks for six months."

Denton didn't respond.

Or maybe he wasn't entirely capable of reading subtext quite yet.

What Steffy meant, in his roundabout way, was that a schedule change so early in the program, just as the company's executives began telling Wall Street that the initial system would be flying in nine months, required more serious forethought. Bad news like that needed a more delicate spin before it reached the executive level.

Of course, he realized that this first crop of engineers would need a bit more time to come up to speed than he had. At least he had joined the company with a few years of direct experience in the sometimes byzantine business of aerospace. The Pegasus team had been young but, as he was beginning to find out, not quite so young as this.

He assured Denton that everything would be fine. At the moment they only needed to take a breath and think about how to report the news. Not everyone up the chain of command was calibrated to hear about a schedule slip, either.

At the end of their meeting, Denton walked out of Steffy's office with an order to revise the report. Steffy had an agreeable feeling that they would soon come to understand each other. The first schedule slip and significant price increase would remain a secret only between him and his young charge, at least until the right people had a chance to work it over. Hands better than his were needed to package a slip for the executive suite.

After thirteen straight hours in the lab, the easy banter between Rob Denton and Steve Gurney had gained an edge.

"Maybe the problem's in your layout," Gurney said.

"Don't start with me, man," Denton answered. "Your timing's all screwed up."

At ten P.M., Denton, Gurney, Annette Mirantes, and Dan Rittman were still working under high-intensity lights in the lab. Ever since they sat down and typed in their passwords that morning, the Freshouts had entered a no-time zone, working uninterrupted, oblivious to the hour, until periodic calls from a perplexed spouse or roommate began to interrupt around the dinner hour, the usual pattern for a weeknight.

The project manager had gone home, and no one had seen Mike Dobbles for hours.

While Gurney and Denton argued with each other, next to them on the bench, Dan Rittman continued to scrutinize his transmitter. When he wasn't talking to Annette or hunched over his motherboard in stern silence, Dan talked to himself, mumbling at times as if in a semiconscious state.

He stared into his hands.

"A hundred and fifteen kilosamples . . . that's one bit every 1/57.6 seconds, so in real time . . . this is . . ."

Lifting the board up to his face, he eyeballed the handiwork, then touched a few spots to see if they had overheated.

On the first board, sent by the manufacturer during summer, the microcontroller had fried itself due to sloppy soldering. That little mistake, he reminded Annette, had cost the team three weeks of schedule. Now that the computer would not download any of the messages Annette sent, the manufacturer emerged again as prime suspect.

Apparently bored by her partner's mumbling, Annette called her boyfriend, Luis.

"Hey, Dan, do you know how many transistors are on the new Intel 586?"

"Like, 3 million," he said, not looking up.

Annette relayed the information over the phone.

"Dan, are you sure that's not 3.5?" she asked.

"Could be," he muttered.

"Could be," she repeated into the phone. "It's a roundish estimate.... Believe me, Luis, he knows what he's talking about."

"What other facts can I amaze you with?" Dan said, holding his board overhead to examine connections underneath.

Dan set the board back on their workbench and tried to pry the microprocessor out with his fingers.

"If I could just get this thing to . . ."

A sudden snap startled him.

"What the . . . " he said. He let go.

"Hey . . . Mr. Science," Annette said.

"Hey, what! Maybe I should get a 900 number," Dan said, and then to himself: ". . . I soldered that thing myself . . . I have no idea, it's not hot . . . unless I burnt out this chip . . . God, I can't imagine . . ."

Annette decided not to relay another query from her boyfriend. She finished her conversation and set the phone down. "I didn't see any extra solder on those pins," she said.

Dan didn't respond.

She tried again to reach him. "Hey, Dan, do you know that guy Alan Parker?"

"Who, our customer? The head of Orb Inc. guy? That white-haired man?"

"Yeah, let's call him and ask if he knows what to do."

"Yeah, right."

Annette found a set of blue-handled hook-nose pliers and handed them to her partner.

"Here you go, Einstein, try this blue thing," she said.

Grasping the controller with the hooks, Dan gave one abrupt tug and the chip popped out.

"Whoa!" he said. "Hey, handy tool."

She retrieved a large set of design pages, the size of circus posters, picturing the physical layout of the transmitter. Flipping to the section that showed where dozens of connector pins on the computer fit into the board, the two engineers scanned the device to make sure all the pins had been fitted into the right holes.

"Something's strange here," Dan said, flipping the board slowly over and over in his hands. He cocked his head back with his face toward the ceiling, his nose almost touched the bottom of the board.

Annette laughed. "Hey, Quasimodo, you just need a little drool out the side of your mouth."

Dan set the board down and shook his head in disbelief. "Guess what?" he said. "I think they put this thing on backward."

Annette gasped. He had to be joking.

"You mean, we pay a professional vendor thousands of dollars to build this thing, and some clown snaps the controller in cockeyed?"

"Yep." Dan repositioned the controller and popped it back in.

They had lost hours to the vendor's sloppy oversight.

She started the test again, firing packets across the serial line.

Dan sighed wearily. "What time do you have to get out of here, anyway?"

Annette kept tapping her keyboard to see if there was any feedback from Dan's board.

Nothing yet. They decided to wait another few minutes.

"So," Dan said, "I haven't met many of the Elephants."

The Freshouts referred to the executives in the company as Elephants.

"What's this guy Antonio Elias like? He designed Pegasus, right?"

They both sat back and chatted. Dan set his feet up on a chair.

This *guy* Antonio Elias was the company's premier engineer, a former MIT professor, father of four, often referred to as Orbital's creative genius. He had come up with the idea for Pegasus one day while grocery shopping—that was the story, anyway. Annette and Gurney had met him a few days earlier to discuss problems with the satellites' operating system software.

"I mean, is he just another one of these guys who knows everything?" Dan asked.

"Not really. He started out the meeting saying, 'I don't know anything about operating systems,' but then you could tell he really knew his shit from all the questions he asked."

"Steffy's the same way," Dan said.

"Right about that," Annette said. "Remember when they were doing those communication tests and he sent everybody home early so he could quit pretending to be a manager and just get his hands dirty? He's an engineer in his soul."

"Yeah, it's funny, he's always going 'I don't know anything about com stuff'—"

"And then he goes off—"

"Fucking shit, he goes off!"

"But that's always about hardware. When he says he doesn't know anything about software, I believe that."

Suddenly Dan saw Annette's screen rapidly scrolling hieroglyphics. She spun around to look.

"This is weird," Dan said.

The screen ran up garbage, then displayed the reception of hundreds of interrupts.

At that moment, on the next bench over, after an entire day's effort, Denton's PC finally posted a message from Gurney.

"Hello!" Denton said.

"Did it say hello twice?" Gurney asked quickly.

"Once."

Gurney smacked his hands together and groaned.

"Okay, it's a software problem," Denton said, with obvious delight. "I knew it. You've got some kind of insidious bug."

"Turn off all the interrupts," Gurney said.

"I'm convinced it's not my hardware."

"Shut up," Gurney said.

Like Dan, Gurney's eyes glazed over, and he began to talk to himself, to move around their bank of electronics, graphic analyzers, and PCs quickly preparing the setup for another test. "I'm going into the operating system—check the ports—someone probably changed some damn thing in this code . . ."

Like sifting through hot sand for a lost contact lens, Gurney spread his arms across the workbench, eyes darting, fingers alive, tapping keyboards, scrolling through lines of code on the satellite's homegrown operating system.

The operating system, a new, never-tested, dense stream of instructions, was known as OSX, but playfully referred to by its inventors in Boulder, Colorado, as "DOOM," a name that had struck everyone as only laughable until recently, when Gurney and Annette began to suspect some rather worrisome design flaws of its own.

"The timing's screwed up," Gurney said.

"What?" Denton replied. "It worked under the old version of OSX."

"Unbelievable," Gurney said, squinting at row after row of tiny gray characters on a black screen.

"Looks ugly," Denton said.

"OSX," Gurney said wearily. "Annette and I told them to trash this thing."

"Do we have to change the operating system?" Denton asked. "I mean, is the message coming in on the right path, or what? I could try tracing it with the scope."

"Good luck," Gurney said, and immediately began writing instructions for the system to recompile.

Gurney tapped "enter," then reached for the phone to call Morgan Jones, their operating system expert, at home. Morgan was the only in-house authority on OSX.

When he finished the conversation, Gurney had a pinched look on his face.

"He says we've got to allocate 300 bytes of memory first," Gurney said.

"What? Is this documented anywhere?" Denton asked.

"That's what I just asked Morgan. He says we're paying the guy who sold it to us to do that, but right now there's nothing. The system's totally undocumented. Maybe in six or eight months . . ."

Denton laughed. "Six or eight months! What's it mean to us right now?"

"It means we've been trying to use a program all day that was never really ready to work," Gurney said, shaking his head.

Denton groaned. "Jesus."

Gurney said, "I guess we could check it once more on the oscilloscope—"

Denton spun in his chair.

"It was sitting right here. . . . Oops, someone's 'borrowed' it."

At the next table, chattering back and forth, Annette faced into her computer screen, and Dan had his head down in their scope.

They were having the same problem.

Gurney tapped Annette on the shoulder and passed on the news about the operating system glitch.

"Man!" Annette said. "Day wasted!"

At a few minutes after eleven, Denton reached out to turn off the bright lights in the lab.

But Gurney was already downstairs, out the front door, heels happily clacking along the sidewalk around the dark parking lot. The lot wasn't as packed with cars as at eight that morning, but still a pretty good crowd. Engineers from the other development projects worked late at Orbital, too. It was the culture.

He thought about calling his roommate to meet him for a beer, or maybe he would just drive into the city and shoot a few games at Buffalo Billiards. All in all, a decent day—still no worries, no fear—and he even had time left to play. He climbed into his Saab and gunned the motor.

Weird name for an operating system, though. Why, he wondered, would anyone call it DOOM?

One day late in September, the discovery of a skylight on the roof encouraged the youngsters to knock out a few ceiling slats in the second floor lab, allowing the lab to flood with warm, natural light. Long fluorescent bulbs recessed in the ceiling also lit the room with an unwaveringly intense light. The effect was heightened by reflections off the white walls, the white floor, the white tables, even their white network computers.

More like an operating room than a workshop, the lab looked so bright that some mornings it required an effort just to adjust the eyes. Even in the evenings, when a few engineers continued laboring over their boards until well after midnight, every hour looked like midday. With no clocks, no calendars, no marker boards inked over with schedules, sometimes an engineer would finally look up after four or five hours of intense work and ask if anyone wanted to take a lunch break, only to realize that lunch had passed hours before.

It was just as they wanted it: intensely bright, luminous at any hour.

The first satellite body, a hard, green, lightweight ring with six-inch-high walls called an engineering development unit (EDU), had come up from the high bay. It looked like a bass drum set on its side and rested atop a set of tables butted together in the middle of the room waiting to be stuffed with electronics. The electronics—individual circuit boards for attitude control, power, internal messaging, transmitting, and receiving—remained unconnected and in various stages of design until that moment when they would cross the first hurdle of testing, then go directly into metallic housings that the engineers would set inside the satellite body with the careful alignment of precision clockmakers.

Still, on most days the lab looked like Christmas morning after presents had been unwrapped. A perplexing variety of tools and materials sloppily stashed away or mislaid—cables, soldering irons, computer manuals, spools of wire, sensors, glue, screwdrivers, pliers, razors, and snippers—found their way out of storage drawers and never quite seemed to find their way back. Head-high test racks, spectrum analyzers, electronic scopes and a half-dozen new Gateway computers created banks of the best equipment to examine the satellite body. The lab was a mess, though essentially a pleasant one.

Bit streams circulated and electrical pulses fired across various subsystems, pushing the temperature of the room from a comfortable 65 degrees at eight o'clock, when most people arrived at work, to about 90 by midafternoon. The noise of unrelated conversations and a roaring pedestal fan set in the doorway enhanced the festival-like conditions. They became aware of a constant, high-pitched buzz, like the noise of electronic crickets, rising and waning as they worked.

Whenever the project manager walked in to look at a new antenna model or pick up a motherboard or just listen to the conversations in the lab, he was tempted to stay.

Dave Steffy would put his fingers on the goods and suddenly realize that he was an intrusion.

At the far corner of the lab, he would hear Denton shout: "You're farting garbage, Gurney!"

Heads would turn.

Noisy conversations swelled throughout the lab.

Steffy would notice one of the technicians mumbling loudly, apparently to himself.

Annette: groaning over her failure to pass a "hello" message to the receiver, sat at her workbench banging the table, "Damn! . . . Damn!"

Grace: arguing on the phone with someone about changes in her contract specifications for the attitude-control electronics.

Above it all: Denton laughing at his partner.

"It's a hardware problem," Gurney kept saying. "I'm telling you, it's your hardware!"

Life in the lab looked and sounded normal, Steffy would think.

In many ways conditions could not have been better for those who only wanted a chance to prove themselves. Whatever his engineers needed for making progress, he would try to furnish. Any needs the company could not fulfill, he and his Freshouts would freely seek to satisfy by improvising on their own, no matter if it was more hardware, software, lab equipment, or as was becoming clearer every day, a few more months of schedule.

Retreating to his office, Steffy would return to his schedules knowing at this stage his team needed only to work without interference from him or anyone else nosing around the lab and peering over their shoulders.

Every day he would visit the bright lab, then turn to leave them to their own devices, even as the din kept rising like a brawl in a smoky bar.

Bulletin Board

"I always felt I could do something that someday might make a difference in the world."

—Mike Dobbles
Orbcomm satellite engineer
June 1992

"Will we bump schedule? I can neither confirm nor deny. It gets very political. I'm not going to sign up to a July '93 launch right now, but we have to tell them something that will sound reasonable."

—Rob Denton
Orbcomm satellite engineer
July 1992

"The whole 'Space Age' thing is kind of funny, right? Putting a man on the moon? That had nothing to do with science, you know. Or making money. Now that we have some real experience under our belts, don't you think it's about time we did something for real?"

—Grace Chang
Orbcomm satellite engineer
September 1992

Internal Design Review (September 1992)

Mission Design Requirements

Orbit altitude: 785 km (nominal)

Inclination: Near polar

Base program: 3 orbit planes with 8 spacecraft each

Inclination: 45 degrees

Separation between orbit planes: 135 degrees

Option: Additional plane with 8 spacecraft

Spacecraft Bus Requirements (January 1992)

Four-Year Life

Subscriber Links:

137.0–138.0 MHz downlink, 4800 bps

148.0–149.0 MHz uplink 1200/2400 bps

Gateway Links:

137.05 MHz downlink, 57,600 bps

148.0–149.0 MHz uplink 57,600 bps

Stacked Disk Configuration:

- flat solar array; 41-inch diameter disk; 6 inches high
- 50-watt orbit-average power capability, end of life

Antenna/communications:

- VHF crossed dipole
- arrays stow onto back of disk, deploy passively

Attitude Control:

Nadir pointing, yaw steering

Attitude control and knowledge: ±5 degrees

- actuators: magnetic torquers only
- sensors: 3-axis magnetometer, sun sensor, and detectors

Mechanical:

Self-deploying antenna

Spring-loaded hinges deploy all elements simultaneously

Foldable mast

Mast elements: 3/16" thick aramid fiber/phenolic honeycomb and Kevlar face skins

Hinge elements: Carpenter (heritage air force meteorological satellite)

"Orb Inc. went out and looked at every possible allocation in this band and acted like they were all operating at the same instant, all the time—the worst possible scenario. Then they calculated all the other kinds of interference in the atmosphere and finally came up with a number that tells us an interference level. And the number is huge. Jan King's been flying satellites for twenty years, and he says he's never see anything like this within a factor of 100. So was it purely a business decision? It's really a cover-your-ass decision so they can point a finger at us if there's a problem."

—Eric Copeland
Satellite engineer
September 1992

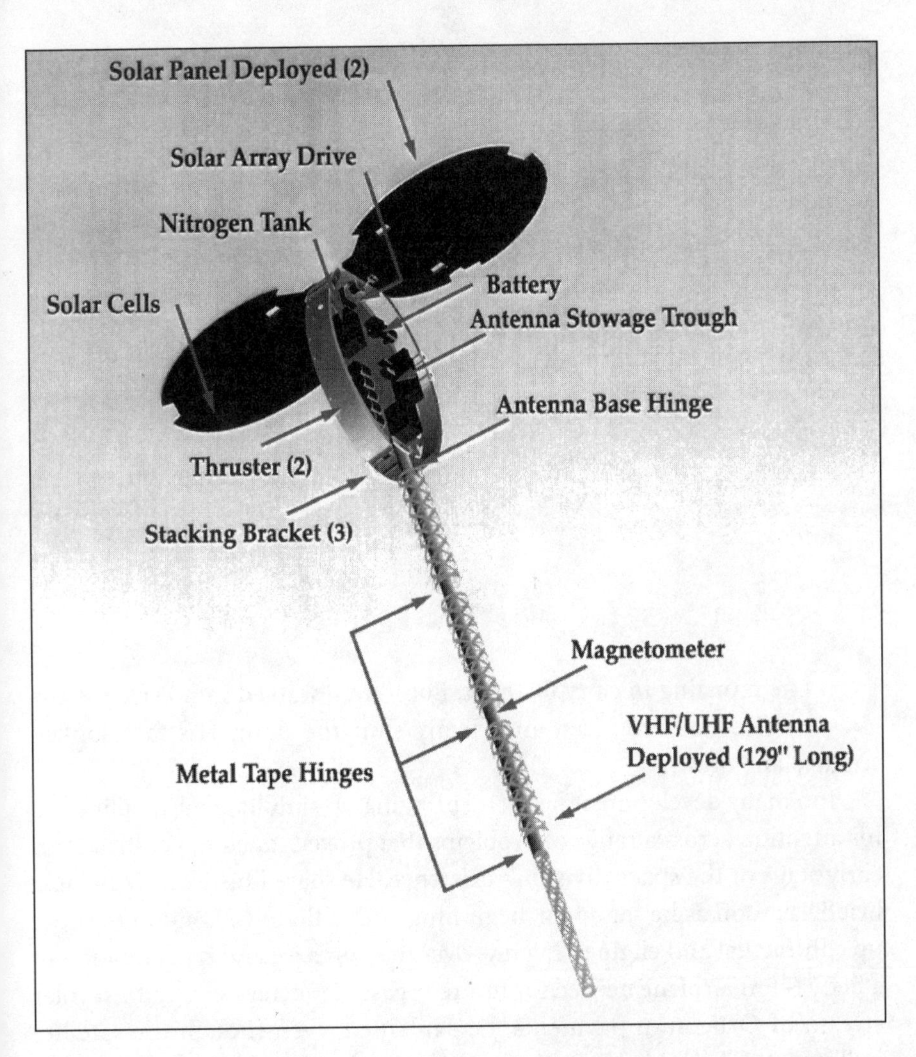

"I have the feeling we're trying to do too much with too little. If we are really able to do it like this, someone would have done it twenty years ago."

—Gregg Burgess
Satellite systems engineer
September 1992

"If you start at zero, there's a lot of room to grow."

—David Thompson
City Paper
December 4, 1992

CHAPTER FOUR

October 1992

Intramurals

One morning in early October, Bob Lovell walked into Dave Steffy's office unannounced and firmly shut the door. His face looked strained and drawn.

Too many development projects springing up simultaneously splintered his attention across a range of problems that proved much too costly at this early stage of the space division's existence. He shared his week's schedule briefly as soon as he sat down, beginning with a flight to London to shore up contractual and engineering nits over the purchase and rehabilitation of a hefty L-1011 airplane needed for future Pegasus launches. Without a regular stream of Orbcomm payments, he explained, even though the satellite builders were falling behind schedule, the company would have trouble purchasing the L-1011.

Obviously, news that Steffy's team would slip the July '93 launch date had serious repercussions. Timing, after all, was critical.

Looking across the top of his spectacles, Lovell said; "Sometimes engineers have technical strengths and miss the business impact, Dave. And that can get you in a ton of trouble."

Lovell was at least twenty years older than Steffy, a handsome, plainspoken, rough-hewn gentleman with impeccable credentials and excellent connections. During his career at NASA, he had coordinated the development of the world's first international satellite search-and-rescue system, known as COSPAS/SARSAT, and bid the first contracts on technological experiments that became the basis for Motorola's Iridium constellation, now emerging, ironically, as Orbcomm's chief competitor. Lovell had brought the idea of the Orbcomm system to Orbital, based on his COSPAS/SARSAT technology, and he had worked side by side with Steffy and David Thompson to design and launch the first Pegasus rocket in preparation for the Orbcomm mission. He was also Steffy's chief proponent in the company.

It was said that both Steffy and Lovell had "good dirt under their fingernails," an engineer's sort of praise, suggesting the men would sooner handle a socket wrench than a flow chart. Steffy, who had designed the Orbcomm satellite at his breakfast table, was the kind of guy who could cut a length of electric cord to a millimeter of perfect by using only the span of his fingers as a rule. Lovell had spent his earliest days in aerospace building nuclear rockets, slogging through high bays ankle-deep in foul runoff fit for fish kills, and he still enjoyed nothing better than mating a satellite to the nose cone of a rocket.

But Steffy's respect for him was not inhibiting.

"I always thought the company could manage a six-month delay," Steffy argued.

"Not if we launch one Pegasus every month next year," Lovell said. After reviewing the division's budget and manifest of launches, Lovell also had concluded that the Pegasus teams needed to grow to meet the backlogged demand for rockets. Again, progress payments for the Orbcomm satellites were essential to offset those costs.

"A launch a month—that's real?" Steffy asked. The company's boast about "a launch a month" had sounded to him like hype from the Pegasus publicity package. Otherwise, he couldn't pretend to understand the division's financial juggling.

"Reality is, we never know what Big Brother wants from us at the end of each quarter."

"Oh! So then that's what this is about—we have to look good on the next quarterly report. You're worried about what stockholders will think."

The boss looked annoyed. "The top priority right now, Dave, is making our customer contracts work. That means making rockets that fly. We have to keep the Air Force happy, and we have to reduce the number of anomalies we've had."

"Yeah, but I thought the third quarter was looking very good."

"Don't kid yourself. We haven't even begun to establish ourselves in the market yet. And to do that, we've had to sell ourselves at prices a little lower than we'd like to."

"Leaving Pegasus with a launch a month? So you lower the bids on Pegasus and screw us. Good luck."

"We all live with tight schedules, Dave."

"Meaning what? We should tell David Thompson what he wants to hear? Or do we tell him the truth—there's no slack in the new schedule—and then just let him try to beat us back?"

Lovell sniffed, a private laugh that could have been interpreted as condescending. "It's all about managing expectations, Dave."

Besides pressure from the top, Lovell also worried that the in-house customer, Orbcomm Inc., would use the schedule slip to fight Thompson's decision to build satellites with Steffy's team. Orbcomm's president had threatened to bid contracts out of house all along, claiming he could find a less expensive and more professional team of engineers elsewhere, at Lockheed or Hughes. He often complained that he was too busy developing handheld communicators, marketing the service, and setting up the worldwide business to fret over Orbital's inexperienced den of wolf puppies.

As Lovell voiced a litany of concerns, Steffy finally interrupted and said he always considered threats from Orbcomm Inc. typical internecine squabbling—the usual "intramurals." As far as he knew, even Lovell might have been bluffing him at that very minute.

The schedule slip represents only half the problem, Lovell said. A threefold increase in the price tag, bumping the cost of satellites to Orbcomm Inc. from $600,000 a pair to a million a piece, really could jeopardize the satellite project. Since their relationship with Orb Inc. relied on a simple handshake agreement rather than a signed contract , the customer might just use the escalation in costs to convince Thompson that they would do better by putting the project out for bid.

"All this bad news is bound to make them feel snake-bit," Lovell concluded.

"Just a minute, Bob!" Steffy said. "I didn't escalate those figures! Orbcomm insisted we add new solar panels and double the performance six months after we settled on a design." The new costs, he added, reflected major changes that Orb Inc. had insisted on.

But Lovell was done, out of his seat, standing at the doorway. An irksome smile rose on his face.

"This project's going to be a shoot-out, Dave," he said.

"Wait a minute!" Steffy said.

"It's gonna raise your blood pressure, for sure."

Lovell had his hand on the doorknob. The expression on his face hadn't changed.

"Look, Dave," he said. The fatherly tone that crept into his voice suddenly set off the project manager.

"Everybody knew eighteen months was a tough schedule!" Steffy stormed. "Nobody in the business launches a new satellite in less than three years!"

"You've got a great project here, Dave. You're about to grab the last big cookie of satellite communications."

Steffy laughed—"the last big cookie." It began a familiar speech that he had heard too many times already. The "last big cookie" lecture always marked the end to any further conversation. It simply meant Lovell did not want to say what should have already been understood between them: . . . *and you better grab that cookie before anyone else slips a hand into the jar.*

The man walked out, and Steffy lowered his head, pinched the bridge of his nose, and shut his eyes.

Less than a year after they had begun, the satellite engineers already languished months behind. At Orbital, under the progressive management of David Thompson and Bob Lovell, pressures on his scrappy little team would only increase.

For many years understatement and secrecy typified the aerospace business in conservative northern Virginia. Projects funded by "black" budgets thrived just beneath the public's awareness at nearby Tyson's Corner or a few miles away in McLean, where the CIA made its headquarters. The same deep silence surrounded an enterprise right across the street from Sullyfield, where a mysterious division of the federal government known as the National Reconnaissance Office had built a multimillion-dollar headquarters without informing Congress or making its very existence public. Spooky satellite outfits with anonymous names like DSI or DRG or IDA not only dotted the Washington Beltway but for years produced space technologies that made the Orbcomm mission look like sandbox play.

By comparison, in the early 1990s, proponents of space commerce began to sound off like blathering puppets on an operatic stage. Relative to spook outfits—spread in wealthy proliferation, snuggled securely in among the shopping centers and commercial districts around D.C.—the new class of commercial space companies was absolutely boisterous.

A startling change, a half dozen small-sat entrepreneurs and constellation developers like Orbital suddenly started banging a drum with public pronouncements about their whereabouts and boasted grand dreams to build rockets and satellites and commercial launch sites for the twenty-first century. Northern Virginia, they said, would soon become to satellites what

Hollywood was to the film industry, what Detroit once was to automobiles. It was a Klondike gold rush.

If the stake was in space, the gold itself was spectrum, that finite range of electromagnetic waves that the world depends on for radio frequency communications. Like any precious natural resource, the use of spectrum is tightly circumscribed by national and international laws and procedures. While half of spectrum use in the United States is claimed by the government, mostly for national security, and huge blocks serve television and radio broadcasting, occasionally a few bands that remain unallocated come up for sale or auction or public use. In early 1992 delegates from the United States attended the World Adminstrative Radio Conference in Torremolinos, Spain, requesting that transmission bands for satellite uplinks and downlinks be set aside for mobile communications, specifically for simple low-Earth orbiting systems like Orbcomm. After stormy debate the international group made a historic decision and agreed to set aside spectrum for data-only systems, known as "little LEOs." It then fell to the Federal Communications Commission to determine how to divvy up various bands in the United States.

At FCC headquarters in downtown Washington, all kinds of companies were filing requests for bandwidth. Some demanded free allocations, arguing that they had developed one astonishingly new technology or another, making them "pioneers" in the field of LEO communications and therefore deserving special preference. Others, like the executives at Orbcomm, sought an "experimental" license and argued that their technology would be so inexpensive and of such general benefit to much of the world that they, too, deserved a free allocation of spectrum. A few companies simply appeared out of nowhere, looking for a piece of the action: schemers funded by the French government, commercial outfits parading as altruistic nonprofit health organizations, "corporations" that existed only on someone's hard disk, lawyers who duplicated other companies' technical documentation and filed it with the FCC under a separate name. As in almost any industry, the satellite business created its own style of hucksterism and chicanery, and the entire escapade sometimes began to take on qualities of a circus.

Although Orbital succeeded in persuading the field of delegates in Spain to set aside a small slice of spectrum for Orbcomm's service, there was a notable lack of chest-thumping heard from the executive suite or public relations staff afterward. At Orbital, executives had learned to be much more understated about their intentions. The Pegasus rocket had generated enormous publicity for the company, and it had not taken long, in the wake of its initial success, for every engineer in the company to realize notoriety exacted a price. Competing in the highly visible arena of American rocketry, they had learned to be discreet.

For instance, whether one of its rockets nailed orbit, missed a trajectory, or veered off course, Orbital could not avoid the keen interest of Washington newspapers. News photographers snapped pictures of launches and reporters wrote stories that rarely failed to mention each corresponding rise or dip in the company's stock price. Newspaper publicity had proved to be a mixed blessing, especially in 1991, when a few unfortunate launches put the company under the hot lens of one of the nation's most influential morning business pages.

After one launch, when the Pegasus placed seven defense-related small satellites in a lower orbit than intended, reducing their effective lifetimes from three years to one, *The Washington Post* zoomed in on the big news—*hometown upstart bungles multimillion-dollar government mission*—and left Orbital's reputation decidedly tarnished.

On another occasion a small suborbital rocket built by the company, carrying Star Wars experiments, went out of control seconds after launch, veered, and slammed into a remote area of an air force station in Cape Canaveral. The brilliant, surging tower of sparks and smoke produced a dramatic scene for photographers shooting for newspapers from Titusville, Florida, to San Francisco, California, embarrassing Orbital's executive suite and raising inevitable questions about whether the company could overcome challenges set by regular production schedules.

By 1992 the company's launch record could have been called a moderate success—fourteen of sixteen launches nailed orbits, matching the industry-wide average. David Thompson put the few failures in context, saying, "If we were a baseball team and our record translated into a batting average, we would be in the Hall of Fame." But questions still arose in the government and throughout the industry about whether Orbital could emerge as something other than an innovative bootstrap operation. The media's attention continually magnified the corporation's growing pains.

By the time the Orbcomm project began, it was apparent the new satellite program could not afford the same degree of scrutiny from the press. There were too many other lions in the jungle. Commercial space was still far too speculative for grand pronouncements, and risks remained too numerous to invite the press in to feed. Although the term *personal communication services* (*PCS*) had already entered the acronym-numbed language of aerospace, thirty-five years after Sputnik the hazards of the frontier were so well defined and the competition so fierce from other Earth-bound technologies like fiber optics and cellular radio that no one at Orbital spoke publicly in absolute terms. Orbcomm simply could not afford constant attention from ill-informed media.

Although the parent company still enjoyed the reputation as one of the fastest-growing in the United Sates, Orbcomm executives made sure their little project remained mostly out of public view, almost as withdrawn as those dozens of other secret, government-related, DOD-financed aerospace projects under way around the Beltway.

The fact was, most American satellites had sprung to life cloaked under classified or undisclosed federal budgets, launched at odd times from remote sites, designed in places few people knew existed. Consequently, over the course of aerospace history, the public never had a chance to observe the industry closely enough to comprehend the trial-and-error nature of launch vehicles or to grow accustomed to actual failure rates among satellites. At the same time, after forty years in business, the nation's aerospace industry, including Orbital Sciences, also remained largely untutored by lessons of the marketplace.

Orbcomm executives realized that a botched launch or distressed satellite would be painful enough without the burden of excessive public attention. Almost no one went into business the way Orbital proposed, sending their own satellites into space using their own rockets, with the backing of private money. As a result, the company was never far from the watchful eyes of *The Washington Post* and Wall Street anyway. Like chip makers and biotech start-ups, Orbital had invited public scrutiny quite unlike any company in the business because of its unusual entrepreneurial nature and ambitions.

At the same time commercial pressures, particularly the response of private investors and Wall Street, were critical. Deadlines and schedules proved to be more important to the company's privately funded ventures than they ever had for projects under government sponsorship. Not surprisingly, competitors, investors, and financial analysts tuned in to each word. Every quarterly report, every launch, every news release had the potential to send ripples of anxiety into brokers' offices and investment firms. Public confidence mattered more than anyone might have realized. As a result, the need to manage expectations factored into every tactical decision, and the need to handle bad news required exceptional finesse up and down the line.

"Put on the gloves!" Alan Parker barked. "C'mon, Lovell! Step up! I'll take you anytime."

The Orbcomm staff watched the boss shadowbox across the conference room. Red suspenders tugged tautly across his round shoulders when he jabbed with the left, slacked when he sent uppercuts into the imagined belly. Standing over six feet, handsome, silver-haired, weighing close to 200 pounds, Alan moved like a meat packer who looked like he had been packaged himself that morning by Neiman Marcus. Shadowboxing suited him—even in silver cufflinks and a lean, black pinstriped suit—especially when his opponent was a middle-aged colleague whose office was ten miles down the road.

"You're gonna win that one, Alan!" someone called out.

"You bet I'm going to win!" Alan said, chuckling with pleasure after the first flurry.

It was late October, only a few weeks after Lovell tried to force Steffy to stick to a July launch. As it turned out, even Lovell had not been able to halt a slip, and news of the new October '93 launch date reached "the customer" quickly.

Alan Parker, president of Orbcomm Inc., had read the memo that morning in his office and exploded with laughter. He was still laughing when he walked into the conference room and pretended to stuff his fists into boxing gloves.

"What a snake," Parker said. "Did you read this memo, David? The head and tail are the same."

David Schoen, Orbcomm's chief technician and the satellite system's designer, said he had.

"We always said we'd never get a decent bid out of house if people knew we were considering Steffy's team," Schoen said. "But without competition we'll never get Steffy's people to sharpen their pencils."

"And it used to be they wanted $300,000 for each satellite," Parker sneered.

"I guess that means they just pissed away the twenty mil we set aside for the start-up," Schoen said.

Parker laughed and jabbed again; Schoen scanned the papers.

"What happened?" Schoen asked. "They've got the same vendors. Everyone signed the same agreements."

"Not exactly," Parker said. "Look at the estimates—they suddenly get higher when they reach the Rockies."

"The Rockies" referred to a distant offsite operation that Orbital maintained in Boulder, Colorado, where a separate team of veteran engineers had Orbcomm's satellite communications equipment in its first stage of development.

"We'd be a lot better off if we had the Orbcomm-X data," Schoen said.

Orbcomm-X. The name triggered reactions around the room.

"Yeah, we'd all be in better shape if they hadn't lost that satellite," Parker said.

Schoen's face reddened. "Every time we have an outside meeting, we spend the first forty-five minutes explaining why we had that failure. We feature the fact that we're becoming one of the few fully integrated space companies in the world, and brag about how we'll build these satellites in record time, but whenever anybody probes a little deeper, we always have that ghost to fight."

"It kneecapped us," Parker said. "What can I tell you."

This was the first reference I had heard in months to the failed satellite. Although I did know about the company's first experimental satellite, built in Boulder before Steffy became program manager and launched unsuccessfully a year before, mention of the satellite never cropped up in Bob Lovell's offices. But throughout the morning meeting at Orbcomm, its name continued to intrude on conversations. Despite a wealth of good news from the marketing and engineering staffs—about sustained queries from Ford Motor, about sudden advances in the network system's software design—the legacy of the aborted Orbcomm-X and its distant creators in Boulder, Colorado, kept insinuating itself ghoulishly into the meeting. The topic of its death and the new schedule overshadowed every item of good news.

"I'm not going to eat this!" Parker said finally. "Like it or not, we're the customer. I'm calling a meeting between Lovell and Thompson. We won't start this project pleading for mercy! That's mandatory! 'Be there!' I want them to feel the heat! I want every engineer in Dave Steffy's lab to feel it in his guts. They can either sign a contract or we take this business out of house!"

At that moment Orbcomm's international marketing director could be heard outside in the hallway sending a fax to the Ivory Coast. Just as the walls started to talk, Parker seemed to sense something foul intruding on his ambitions.

"If I see another memo like this," he said, "I'll march down to David Thompson's office and say, 'Stop the program.' We're not a giant company. We don't have a lot of options. But I do not want to be associated with a failure! And neither does he!"

Orbcomm-X, schedule slips, failure—intimately linked in his mind.

For some reason Parker also acted as if he suspected the satellite builders, having handed him one loss, might undermine him again. The evidence wasn't entirely clear, but I could tell that any mention of the ghostly satellite or of the offsite team of satellite engineers in Boulder made Mr. Parker recoil. Just the thought of them made him act like a most bedeviled man.

If Orbital's problem was to build the first two satellites in twelve to eighteen months, launch them, test the system, and start production on at least twenty-four more, Alan Parker faced an even more awesome set of hurdles.

He had to persuade one or more corporate investors to contribute more than $100 million to help fund the development project and to devote corporate—hopefully, international—resources to inaugerate two dozen satellite franchises in countries around the world. He had to steer regulatory negotiations at the FCC that would encourage a quick licensing process and win special status for an experimental license. Between flights in and out of Dulles airport, while he met with executives at Panasonic in Japan who wanted to build the Orby communicators and traveled to Spain to seek delegates to support worldwide authorization of spectrum for his system, he also found himself fighting with Bob Lovell about increased costs and the lagging schedule. While the pressures made him anxious, he never shrank from the work. In fact, Parker seemed to excel most often when he found an occasion to scrap.

"I'm going to strap you fellows to dynamite," he threatened with a laugh when he first called Steffy and Lovell to sit down to discuss their contract.

A sharp bellicose streak disarmed competitors and amused his closest colleagues. Chiding his own staff, Parker would growl with delight at staff meetings when he demanded of them, "Arm in the fire, boys!"

For those on his side, Parker's outbursts were amusing, even invigorating, but outside his staff, especially at Steffy's end, he presented a threatening, adversarial force.

After all, Parker had learned the game in places much tougher than Orbital. Formerly chief executive of the Ford Aerospace Satellite Services Corporation, vice president of Ford Aerospace, and an executive at Ford Motor Company for twenty-five years, he had come to Orbital in 1989 at age fifty, a veteran of international marketing wars. Although he had left Ford after the company decided to sell its aerospace division, there was nothing bowed or beaten about him, nothing soft or aged, either physically or emotionally. A strong-willed bull of a man, he had rebounded after his career to root out consultant jobs around the Washington area. He had been looking for a break that would put him back into the aerospace business when David Thompson called.

Hired as a consultant, Parker initially resisted Thompson's urgings to take on the day-to-day operations of Orbcomm. A cursory review of the system's early business plan had not caught his imagination. The plan, drafted by

Bruce Ferguson and others, would serve a variety of markets, such as monitoring of remote assets, sending emergency messages, tracking ships and trucks. Parker quickly observed that the business was much too mundane for his tastes.

"At my age," he would tell Thompson, "I'm just not interested in a paper route."

But as he conducted his own research, Parker also slowly began to see the system's potential. Dozens of applications, in fact, from emergency messaging to cargo tracking, from industrial applications to communications support for military systems, could deliver a gold mine. At one point he realized that Orbcomm would break even if it operated in the United States alone, and then he thought of the big win if they could break into the consumer automotive market—an Orby in every car.

Approaching former colleagues at Ford, Parker asked if the automotive industry would take an interest in the technology—yes, they would. He went to the American Automobile Association and asked the same question—yes, they would, too. If he expanded the original idea from a two-to-four-satellite system to thirty-six, the market exploded five hundred times the original estimates. "You can break even in the U.S., and all the international revenues come back as pure profit," Parker told Thompson. "This is business nirvana."

Appetite whetted, Parker hired his own consultants and took control. In just a few months he hammered out a strategic plan, then spent the next three years implementing it, often without more than one or two staff people alongside to help.

Boisterous, charismatic, cunning, he boasted that he could bring the business up worldwide within five to ten years, and then began traveling the globe, personally courting international delegates to the 1992 World Administrative Radio Conference, the principal gathering for global spectrum allocation. Once WARC approved spectrum for Orbcomm and its competitors, he turned his attention to the FCC, where federal authorities would decide just which U.S. company deserved its imprimatur for exclusive or shared use of the world's precious frequency bands. Beyond that, he made the initial approach to Ford and AAA and established contacts with dozens of financial investors and corporate honchos whom he thought might sponsor a partnership.

As a player on the Orbital team, Parker's personality contrasted significantly with the quieter, more intellectually persuasive salesmen among Orbital's founders, but his ambitions rivaled their own. Within months of taking charge, he had prompted a study by Ford Motor to evaluate Orbcomm satellite systems as a potential product on Ford's upcoming line of luxury automobiles. He commanded an impressive international campaign that won support of delegates from third-world nations for Orbcomm as a low-cost answer to inadequate communications in isolated and underdeveloped areas of the world. Single-handedly, he delivered crucial votes from the most unpredictable places—Morocco and Ethiopia—and established relationships with key delegates to the world radio conference who promised to speak on his behalf.

Despite sometimes being adversarial, arrogant, and combative, Parker was also a charming and engaging spokesman. He described himself as a marketing man, at heart, and he boasted that he could sell a satellite system the way an automobile dealer sells cars. He would sometimes remark that the satellite business was more like fashion merchandising than engineering and technology innovation. It was, in fact, not much different than selling Fords.

In March 1992 delegates at the world conference in Torremolinos allocated a small sliver of spectrum (4 MHz) for "little LEO" ventures like Orbcomm, and the FCC began to look more seriously at Orbital as a contender for mobile satellite services in the United States. Lawyers for the agency signaled that they would listen to Parker's licensing proposals and consider offering an early experimental license for his first two satellite "prototypes," the satellites that Steffy's team had already begun to build.

By the summer of 1992, swiftly crossing hurdles, Parker found himself racing ahead of rivals like Motorola, Lockheed, TRW, and other more expensive, telephonic "big LEO" ventures that still faced years of regulatory wrangling. People in the industry began to wonder if Parker really might create the world's first commercial global/mobile satellite communications system years ahead of gargantuan corporate contenders.

Inside the corporation it could be said that Parker's entry into the business was more of a takeover than a hire. Years of experience in automotive marketing and international operations at Ford provided him with elite contacts in the telecommunications business and a natural mastery of international business procedures, both of which Orbital generally lacked. There were others in the company who wanted his job, but as Parker secured the business plan, he also began steadily to erect psychological and legal boundaries between the Orbcomm business and the rest of the Orbital corporation. He realized quickly that his status as an "in-house customer" was essentially a weak position, making him dependent on other divisions of the corporation to build and launch his satellites. To take the giant steps necessary to bring Orbcomm to market, he decided to act as if the rest of Orbital were just another enemy to be overcome on a long march toward independence.

The way he belabored every detail of the evolving service, from the satellite's technical specifications to his employees' office floor plan, demonstrated the ferociousness of his need. His insistence on creating legally binding contracts with colleagues inside the corporation and his demands for sizable outlays in financial and human resources slowly helped him enlarge his staff and begin the process of creating a company that would someday declare its independence from Orbital and operate as a subsidiary only. His plan to construct a separate headquarters was another persistent theme he carried like a banner. Every day he arrived at work with what seemed like an infinite hunger that could never quite be satisfied.

The truth was, Alan Parker could not have cared less about satellites or engineering or the glories of aerospace. He would even say so directly and laugh at those whose impulses made them more starry-eyed than he. He called aerospace software engineers "hacker-weenies," and he spoke of satellites as nothing more than a "bent pipe" for passing bit streams. The true value of a satellite system occurred only at the point where those bits could be counted, priced, and sold like crude oil. Money, power, a global business—the chance to provide the world with a great new service—mattered ultimately. His intention was far more pragmatic, much more directed than that of many of his colleagues, directed, as he would say "with laser-beam intensity" at a specific goal: tremendous revenues, $1 billion a year.

"In five years I'll be fifty-eight years old," Parker told me one day, sitting in an office sprinkled with memorabilia of his recent travels in Europe and Asia as well as his long career at Ford. "To me it's very important that I work on something significant because this is probably the last time I'll ever get to do a big project with this kind of energy. I've got four years of my life into it now and I've got five years left. Hopefully, by that time, we'll bring the system up, make a few bucks, and I'll be out on my boat."

Until he reached that boat, though, the satellite business was nothing less than jungle warfare. Beyond his own office walls, Parker saw snakes lurking around corridors and conference rooms. Some were competitors, others colleagues, but in any case he acted as if he expected they were all eagerly waiting for that moment when the big man might stumble or fall.

In just a few years, he would capture the world with one last tremendous win.

He planned to stun them all.

Parker walked out into the hallway after the meeting and joined other Orbital executives who had gathered around the secretary's island to hear the latest news. The company's most lucrative product, the TOS booster, had made its maiden launch that afternoon at four o'clock, separating from the third stage of a Titan rocket, and it was supposedly guiding the Mars Observer on a journey to the red planet.

Unfortunately, after separation, the air force reported, its tracking planes had temporarily lost the signal from the TOS, a predicament similar to what happened to Orbcomm-X.

At the moment no one knew if TOS had failed to fire its rockets or if the booster had nosed down and might be bearing the $500 million satellite on a plunge into the cold Pacific.

Parker snorted and shook his head.

"Can you imagine?" David Schoen said. "What if we worked that project?"

"I'm glad we don't do launches," said one of the marketing guys. "It makes my stomach grind just to think about what they're going through right now."

"They?" Parker said.

All the development projects under way at Orbital at the moment depended on launches like TOS to succeed. Every rocket ride, every contract had to hold up under the glare of publicity. It was sometimes hard to realize, but dozens of other companies around the country were watching Orbital closely, too, many with their own commercial satellite projects gearing up to ascend into low-Earth orbit. Within ten years they would spend billions of dollars to launch hundreds of new satellites into space, each one seeking a niche, gambling on a global market.

The pressure was on, as far as Parker was concerned. Didn't matter if it was TOS or a Pegasus launch—if his staff didn't feel the heat now, maybe he just wasn't pushing them hard enough.

He rubbed his belly, then grinned at his colleagues. "Our day's coming, friends. You'd better believe it!"

CHAPTER FIVE

October 1992

Cowboy Satellite Hackers: Orbital West

Dave Steffy stepped off the morning flight from Washington wearing jeans, tennis shoes, and a flannel shirt, looking every inch a Westerner. Alan Parker's impending contract weighed heavily on his mind, but more importantly, he had compiled a list of "things to do" summarized by the words *antenna, specs, attitude control.* He wouldn't leave Colorado without redirecting the Boulder crew.

A short drive from Denver up Highway 36 left him traveling the outskirts of town. The splendors of bright autumn slowed life to a lovely, relaxed pace. Dappled by the light of golden aspens, traffic on the main roads moderated and gave way to the quiet flow of bicycles. Students from the University of Colorado overwhelmed the town, bringing business a collegiate friendliness and an agreeably easygoing tempo for work of all kinds.

Steffy pulled into a quiet, sunny neighborhood and spotted the Orbital company logo posted at the corner of a piece of property that looked like a modern branch bank. Around the property he could see a hammock stretched beneath two willows and picnic tables set beside a stream that meandered in from the mountains. He parked and walked into the new one-story brick building.

He noted the facility's warmth and cleanliness. Beyond the lobby the central lab spread out from the core like a clean, well-lit cafeteria. Large rows of windows surrounded a common courtyard. The illumination of autumn colors spilled in to bathe the workplace in a cool, dry light.

Steffy had spent a fair amount of time flying the red-eye special from Dulles to Denver. He tried to check in a couple of times a month on his western crew. Generally he regarded it as an enjoyable break from the increasingly jacket-and-tie corporate pressures at home.

Orbital's Virginia headquarters had changed noticeably as it grew that year. The year marked a turning point in the life of the corporation, as Orbital signed a new series of contracts with NASA and the U.S. Air Force for future Pegasus launches, and David Thompson expanded his engineering staffs for rockets and satellites and made plans to increase the size of his corporate staff as well. To fund the maneuvers, J.P. Morgan had led a syndicate of banks to create a $40 million line of credit for the company. A $55 million public convertible debenture offering was created by Lehman Brothers and Alex Brown & Sons. The firm backlog of orders for Orbital products had grown to $230 million.

As a result, the environment at work also became more stressful and more political than ever. Increased business brought more employees into the company, fewer Freshouts, and a new class of veteran engineers with aerospace experience. The experienced hires came eagerly jockeying for midmanagement jobs and prestige projects, creating, in some ways, less pleasant facets to the corporate culture. Steffy found himself competing against three other program managers, as savvy as he, who came to create new projects and sought to expand their financial clout to hire consultants, contract "jobbers," and pay for Orbital's increasingly busy designers and technicians. Steffy was learning to play his cards close to his vest and also, if possible, to avoid entanglements with Alan Parker and David Thompson. Those who knew his game thought he played it very well.

Boulder, on the other hand, seemed as if it existed in a different country entirely, a light-year away from the corporate tensions of northern Virginia. Under the influence of the university, industry succumbed to the pursuit of good health, lush mountain trails, and pleasant intellectual training. Even at Orbital's new million-dollar Rocky Mountain facility, business still reflected the former playful ease and ideals of the parent company.

Engineers at Orbital West recognized "casual day" every day. Formal meetings occurred infrequently. Outside the offices all roads led not to a tollbooth or beltway, but to the remarkable flatirons and ranges that suddenly appeared at the end of residential streets, and to hiking trails that snaked through scrub meadows and rose on rocky terrain overlooking a university that did not come awake until noon. Was there any place so weighted to the ground, so accurately named for its anchor? Boulder seemed an odd place for satellite engineers—a sure distraction from the insistent goals of a growing corporation—or at least so said Steffy's bosses and envious colleagues in Virginia.

Despite criticisms Steffy had spent enough time with the western crew to trust the team's commitment. Boulder's engineers—a self-indulgent and arrogant group, according to their East Coast counterparts—sometimes appeared just as harried, harassed, and eager as the crews back home. They did spend a fair amount of time hiking in the mountains, but they also lived, like all good engineers, under the same strain of monkish austerity. As the support staff for three ongoing satellite projects at Orbital, Boulder's crew had added to their load lately and they also had developed a few projects of their own.

The only real snag was that they had always thought they would be the ones running the Orbcomm business by 1992, not Alan Parker. And, of course, they had that other niggling complication: Orbcomm-X.

Jan King met Steffy in the lab.

Steffy had watched him through the glass of his office talking on the phone, and wondered why the door was closed. When he stepped out, King seemed anxious.

"Lovell just called to say he got a letter last night from Alan Parker," King said, breathlessly. "He's directing us to fly a follow-up mission to Orbcomm-X."

Steffy glanced at his watch. "Boy, he was up early." Actually, he could have guessed the order would reach Boulder sometime during the day.

"Furthermore, they say I'm supposed to make it happen in the next six weeks without affecting any other mission."

Still behind on his own satellites, Steffy needed the Boulder team to push ahead on Orbcomm, and not bog down on a science project for Parker. Steffy set his briefcase on the floor.

"I thought they would give you more slack," he said. "Our communication gear's supposed to be the top priority out here right now."

"No, Alan's jumping on this thing with both feet. Here's the deal: They're giving us six weeks to build two satellites!"

Steffy laughed immediately.

"Get the secretary to put in some overtime," he joked, grabbing his briefcase again.

"Oh, right. Teach her to solder?"

They walked toward King's office.

"So," Steffy said, changing the subject, "how was your ride on the Blackbird?"

I followed along, eavesdropping as they talked.

It turned out to be a coincidence that I arrived the night before. I had come hoping to spend a few days learning more about one of the country's premier figures in small satellite technology—this unaffected, white-haired engineer, Jan King. Formerly an executive with AMSAT, King managed Orbital's Boulder division and supervised a small crew of engineers who had taken on the difficult job of building Orbcomm's satellite communications gear, easily the most challenging part of the mission.

Fortunately, I also managed to arrive in time with Steffy to see the first results of an experimental flight on an SR-71, a manta-shaped Blackbird, that had taken place the previous weekend. Flying more than three times the speed of sound, the "bird" had flown across half the continental United States in about an hour and a half collecting data from existing terrestrial interference in the 148–150 MHz band, which Orbcomm had claimed for its up-link.

As we walked through the lab, King pointed to an object the size of my laptop resting on a workbench. The device had flown on the SR-71, taking data over the weekend. It hardly looked like a satellite, more like a piece of Samsonite luggage. But with some effort King believed he could refurbish it to fly in orbit.

"We'll take that and its spare to upgrade, if that's what Parker wants," King said. "We actually work pretty good in that mode—just take the pieces, throw them on a table and get to work. 'Course he keeps saying, 'You did it in two and a half weeks the last time. You can do it again.' The last time it was Orbcomm-X, or has he forgotten?"

Jan King differed from most Orbital executives. On business trips to Virginia, he gave the impression of being just another friendly country boy, absent the ambitions that typified engineers in other top management posts. He would show up in meetings with Alan Parker and David Thompson, who looked like Armani models, wearing a light-blue cotton suit and a wide, boisterous polyester tie, as dashing as an assistant manager for a neighborhood Radio Shack. He gave formal presentations that strayed with humorous, folksy asides. Sometimes he would try to stifle his perky midwestern twang and iron it out with a more formal tone, but the result always seemed like an imperfect impersonation of someone else.

King was the kind of person whose keenest interests related to esoteric topics—hobbies that combined, for example, a studied knowledge of harmonic frequencies with theoretical possibilities for digital signal processing. He did not have the aplomb of a man like Alan Parker. Engineering, not social manipulation, seemed more his style, a natural way of life, a very singular calling. Outside executive conferences he laughed a lot. He enjoyed discussing technical subjects at length—over lunch, after work, between meetings in the lab. It was clear that this was not a man for whom an indulgent weekend spree would land him at a blackjack table or sailing in the Caribbean, perhaps not even sitting front-row center at the theater or symphony. That morning when Steffy arrived, in fact, King had just returned from a weekend in Death Valley with friends, whose entertainment amounted to a contest bouncing radio signals off the moon.

No engineer in the company quite matched him. Even if they all exhibited one or another of his traits—a passion for gadgets, an intuitive grasp of radio physics—King came across as the company's most wholesome composite, the unpolished prototype, unaffected, unbridled. He was the true article, a quirky, creative guru who had survived the old Space Age by envisioning something else entirely.

Legendary in the league of international satellite operators, he and his team of engineers and technicians numbered less than a dozen in all. But as a group, they could make delicate satellite construction look as simple as a shade-tree mechanic's weekend business. Old aerospace trade publications sometimes pictured him with a team of casually dressed engineers working on spacecraft in someone's basement. If David Thompson's aspiration to build aerospace products "smaller, faster, cheaper" had any existing example, Jan King would probably have been among the few great practitioners in the world.

His engineers had long ago mastered the art of retrofitting aerospace technology for small satellites. By using off-the-shelf components and leftovers acquired from friends in traditional aerospace companies, they had produced a series of lightweight satellites known as Oscars, small machines launched without fanfare for decades under the auspices of an informal consortium of volunteers from around the world. Actually, a number of their products still orbited and functioned well. Like ham radio fans, AMSAT volunteers had managed to turn their love of satellites into a part-time, loosely knit underground profession. As a member of the inner circle of the organization, King helped create and maintain a quaint family of renegade satellite makers in different countries across the world, professional "amateurs" who could be called on to donate labor and material for the "home-brewed" Oscars.

His reputation also rested on the uncanny ability to coax or attract other highly skilled engineers into the private club and to prescribe meticulous production standards in the most domestic of environments—garages, basements, and kitchens. They disdained the industry's dependence on military-rated parts, custom-made pieces that evolved slowly to perfection through exquisitely monitored tests at costs far exceeding the satellites' material value. Whenever possible, the AMSAT crowd relied on commercial parts, standard microelectronics used in the computer or radio industry, off-the-shelf processors, and ordinary electronics, all readily available, mass-produced, and certified in the test bed of massive consumer markets. They often transformed mundane or warehoused materials into humble, inexpensive, long-lasting space machines. The lack of paperwork, oversight, redundant systems, and military-rated parts allowed the AMSAT volunteers a substantial savings over industry-produced hardware.

To build an Oscar satellite, for example, King once acquired a set of solar arrays from an old weather satellite, sawed them off into pieces, and bonded them onto his payload. During the design of one of his last projects before joining Orbital, he had invited a group of Thiokol engineers to his house and, over a bucket of Kentucky Fried Chicken, persuaded them to give him a sophisticated kick-motor, which was collecting dust in storage with a collection of other strategic assets for ballistic missiles. The antithesis of blubbery, government-based missions, AMSAT projects quietly succeeded on shoestring budgets, small, clever designs, a high degree of personal dedication, and a sometimes seedy assortment of found materials.

Legendary skull-sessions over beer and pizza at a roadhouse joint near NASA's Goddard campus in Greenbelt, Maryland, where King once worked, led to the design of Oscar satellites on napkins and paper placemats. He and his engineers used to go home at night after their roadhouse fetes and begin building components on the flat surfaces of rec-room pool tables. They baked epoxied transponders in kitchen ovens.

Between projects, while King lobbied federal agencies to authorize AMSAT to bolt onto air force spacecraft and national meteorological payloads, satellite amateurs used the existing fleet of Oscars in orbit to talk with astronauts on the space shuttle or to tune in to their own exclusive communications links that connected a vast network of stations from southern California to Berlin to Capetown to Tokyo.

For almost fifteen years King and his friends had stowed satellites on the backs of mammoth payloads inside gargantuan, government-financed rockets. They exchanged a few pounds of ballast for the exorbitant cost of tagging along "piggyback" into orbit. Sometimes for as little as $6,000, he and his colleagues performed the miraculous feat of orbiting their own homemade projects. Many of the satellites were still in use in 1992, years after taking orbit. Modest forms of global communications, mapping, and scientific data collection engaged amateurs throughout the world for next to nothing in cost. In a twenty-year career, his amateur coalition had built fourteen spacecraft, as

King would say, "for fun," and launched nine, not counting the half-dozen missions that he had helped develop as a NASA engineer.

Perhaps nowhere else could David Thompson have found a more suitable team to fulfill his ambitions in Microspace.

There was only one problem: When the AMSAT team agreed to join Orbital, King and company had come with an understanding that King would become the Orbcomm program manager and his engineers would design and build the satellites. Alan Parker was supposed to have been his counterpart, and Dave Steffy, his employee.

Unfortunately, after hiring in with a reputation as fine as any in the business, the team suddenly found itself living with the ghost of its first lost satellite. The expectation that they would take over a large portion of the Orbcomm business consquently dissolved. Instead of leading the organization, as they had expected, King and his engineers found themselves working with Freshouts, taking orders from young project managers like Steffy, and having to endure a not-completely-accurate reputation as corporate prima donnas. Parker's contempt rarely reached King directly, but friends in Virginia who had heard his opinions of Boulder's work sometimes told King pointedly, "Watch your back."

Fair or not, the memory of a dead satellite lingers long into an engineer's life. As they say in aerospace, no screwdriver on Earth can fix a broken satellite in orbit. The inestimable costs—particularly the cost to reputations other than one's own—far exceed the millions usually lost in revenues. That one miserable fact alone, the most unforgiving law of aerospace, tainted the formal relationship between King, Parker, and Steffy, haunting them each quietly and in the most understated and venomous of ways.

The Orbcomm-X mission had originated in the early months of 1990, not long after the branch offices opened in Boulder.

In defining the technology, Parker had searched for appropriate portions of spectrum that would suit his business plan—the plan being to make the service as inexpensive as possible. As a data-only system, Orbcomm needed only slender patches of spectrum in the UHF/VHF bands, which would allow the company to rely on commercially available electronics—radio parts, essentially. With the help of consultants, Parker considered several slices of bandwidth. Soon he also realized that interference in any band he selected would have to be well documented so he could assure investors that it would operate clearly without terrestrial noise.

Choosing frequencies was perhaps the most significant risk in the project. At the time most of what was known about radio interference within particular UHF/VHF bands from space was based almost entirely on theoretical assumptions. Although U.S. defense satellites had, in all probability, collected an enormous amount of data on space noise across the spectrum for decades, the federal government prohibited public collection or publication of such data on the grounds that knowledge of those matters would jeopardize military defense. Whatever documentary evidence already existed remained strictly classified.

King's engineers thought they knew better. According to their experience and their sources—most likely other satellite engineers who had worked "black" projects—bands under review by Parker should have been clear enough. If they were not totally free of interference, King believed, they would at least be quiet enough to support a simple data system like Orbcomm.

Parker couldn't be convinced. If the bands he had chosen proved noisy, Orbcomm would be doomed from the first launch. Likewise, if Orbital could not provide data that argued convincingly for specific spectrum selections with the International Telecommunications Union (the worldwide organization that allocates radio spectrum), then Orbcomm would never open for business. The need to choose the clearest set of frequencies weighed so heavily that Parker finally ordered the Boulder team to build him a small satellite. He wanted something simple but elegant that would collect information on interference levels from space in the 32–39 MHz band. Engineers in Boulder went to work on a lightweight satellite the size of a briefcase, which they called Datasat.

In the meantime other consultants hired by Parker began to consider alternative bands they thought might prove more efficient and less costly overall.

Indecision about which frequencies to employ finally stymied King's engineers. Rather than continually rebuild the little Datasat as Parker's consultants made other choices, they simply decided to put the unfinished satellite in cold storage—squirreled it away in an old Frigidaire that one of their technicians had bought at a used appliance store nextdoor for twenty bucks—and await more definite orders.

A few months later, in February 1991, Parker called with a firm set of frequencies—137–139, 148–150 MHz—and said he had scheduled a piggyback ride with a remote-sensing satellite set for launch in July on an Ariane rocket. With only five months to complete the instruments, the Boulder team revived Datasat, renamed it Orbcomm Dos Equis (a bow to the popular Mexican beer), and went to work days, nights, and weekends overhauling for an impending launch.

Orbcomm-X, as it was then officially known, presented a clear expression of classically conservative, awesomely inexpensive AMSAT ideas, reflected most ostensibly by an unusually simple and efficient design. For one, it weighed only 16 kilograms, about 35 pounds. The engineers applied solar panels to each face of the satellite and connected "antennas" made of Stanley tape measures to one side, positioning them to spring out once the satellite ejected from the rocket. The attitude-control system was perfectly passive: the tape-measure antennas, painted black and white on alternate sides, would slowly spin up the satellite once in orbit—arranged like radiometers on the inside of light-bulb toys, the antennas relied on differential moments from light-particle beams bouncing against its black and white blades to force the spin. Engineers placed tiny magnets strategically inside the spacecraft that would interact with magnetic fields of the Earth and stabilize the body after its initial tumble in orbit. To avoid building a frequency scanner from scratch, engineers, also in typical "amateur" fashion, retrofitted a scanner from an inexpensive, commercially available receiver called a "Handy-Talkie," a ham radio implement often used to identify satellite signals from the ground. They stripped the guts out of the "Handy-Talkie" and reconfigured it to make a tunable receiver that monitored two megahertz bandwidths in narrow steps across relevant portions of the spectrum.

Technically, the satellite was a cinch; logistically, the schedule swamped them. From experience the Boulder team knew that, in most cases, launch dates slipped. That was simply a fact of life in aerospace. Having built and launched more than a dozen satellites themselves, they had learned to take any initial schedule, multiply the number of days to launch by 1.7, and fairly predict the actual launch date. A margin of a couple of weeks, in any case, they assumed, would probably bump the date into August or September of '91, giving them time to do an assortment of tests, particularly with the satellite software, which required a "burn-in" time of several hundred hours.

Instead, the launch held firm. By early July the Boulder engineers found themselves scattered across the world, some working night and day in a clean room near the launch site in Kourou, French Guiana, and others at Orbital headquarters in Virginia debating with Parker and other executives the need to forgo an immediate launch so they could continue testing. In the end, rather than risk losing a relatively inexpensive ($200,000) "piggyback" launch or risk damaging the satellite by flying it back home, the engineers crossed their fingers and agreed to launch Orbcomm-X without a final round of software tests.

More than their own reputations were at stake. David Thompson had unveiled the satellite at a public ceremony that spring among a group of dignitaries in Virginia. The state of Virginia had kicked in $250,000 to fund the project and representatives from the Environmental Protection Agency had offered equipment to carry out tests with the satellite. AT&T, Ford Motor Company, the American Automobile Association, and Allstate Insurance each had heard detailed presentations about the Orbcomm system and awaited the results of Orbcomm-X data. The FCC expected to see whether a successful launch would provide data that made Orbital eligible for an unusual "experimental" license status from the agency, recognition that would grant competitive benefits toward the launch of the first two satellites. Competitors, private investors, and Wall Street firms also expressed curiosity about the launch, suggesting that detailed questions would follow as data poured in from space.

"When Orbital's Orbcomm-X satellite goes up this summer, it will carry with it Virginia's hopes for achieving a $10 billion space industry within a decade," Thompson announced in a posh ceremony at the Center for Innovative Technology, a state-owned new-technology think tank. As Orbital's first satellite, Orbcomm-X was heralded like the arrival of a new baby.

Within hours of launch from Kourou, French Guiana, on July 16, 1991, Orbcomm-X nailed its 480-mile circular, nearly polar orbit. But on its second revolution around the Earth, the satellite suddenly let out a loud squelch. One of the Boulder technicians heard the scream on his radio monitor while driving into work. He reached for his tuner and adjusted the volume. Nothing. Pure static.

The satellite had gone mute.

Over the next few hours, every effort to make contact failed. Speechless, perhaps brainless, it seemed to have simply disappeared, essentially vanished, a tiny insectlike corpse trapped in the dark forest of space.

When Orbcomm-X died, it did not fall from orbit. The box seemed fine, at first, as it tumbled. Its transistors fired well enough. From all indications its solar cells were still soaking charges of sunlight and its antenna was still blasting away. From the ground the engineers imagined that its batteries may have still been pumping power, even through the first eclipse. Then suddenly it was gone.

Unlike many lost satellites, Orbcomm-X never left any sign of why it disappeared. Satellites rarely simply abandon the Earth without comment, but sometimes it does happen that way. The industry lived with ghosts of a few nightmarish *blundersats* for which, once launched, Murphy's law ruled. Anything that could happen, probably had: a slim cable harness squiggles loose during launch and snags an antenna when it tries to flip open, leaving it curled fingerlike, swirling about the dead weight of satellite carcass; the avionics software responds to correct torques from a minor solar flare and then enters an endless feedback loop, muttering to itself like a streetcorner moron; a collection of ionized electrons suddenly flips a logic state in the control system, making solar arrays flap like duck wings until the satellite literally beats itself senseless. Just as every nightmare has its own scenario, the death of any satellite often stands out as a unique, perplexing event, escaping interpretation.

On those occasions when the cause defies investigation, an enduring torment attacks its engineers. Not knowing left a residue of doubt that could not be forgotten or, like nightmares of a subconscious sort, analyzed and picked apart. Not knowing was torture, the engineering equivalent of limbo.

After the satellite fell silent, the Boulder team spent weeks testing models of Orbcomm-X in its lab, searching for clues that would explain what happened. They traveled to the National Naval Observatory and used a powerful 30-foot dish antenna to blast signals at places in the sky where they predicted the satellite might appear. Every effort failed. They never again made contact with Orbcomm-X. Within a few weeks they couldn't even approximate its position in the sky.

Fortunately, Parker had arranged enough support among delegates from around the world to avoid a technical challenge with the ITU. The Orb-X failure did not prevent him from winning the frequencies requested. Politically, Parker had played his cards very well.

At the same time, Orbcomm's potential investors turned cold with news of the failure, a response that proved almost as disastrous as having lost the bid for spectrum.

Parker made it known that he would never completely trust the Boulder crew again. Especially among his own staff, he rarely hid his disdain.

By 1992, as the first contracts between Orbcomm and the Space Systems Division were being written, the depth of Parker's distress became apparent to everyone in the company who might have cared, particularly to Orbital's talented but increasingly dispirited team of aerospace engineers in Colorado, who were now part of a new, inexperienced crew of engineers charged with building the *next* spacecraft.

Most dealings between Parker and King quickly turned combative.

King shuffled through piles of documents on his desk and retrieved a tablet of computer printouts. He flipped through pages of distinctive graphs whose plots of data points looked like one continuous, poorly cropped hedgerow.

Whenever he came across an unclipped strand of the "hedge" swinging dramatically above the rest, he paused to scratch a note in pencil.

"The X-axis here tells you how busy the channels are," he explained to Steffy. "The vertical axis—Y—is the uplink frequency range from bottom to top—148 to 150 MHz."

The system had been designed to share VHF frequencies with other existing services, such as paging systems and military radio, by using only 898 KHz of bandwidth.

Conceptually, it was simple. An Orbcomm satellite would transmit a beacon to anyone in range telling which channels in the frequency band were open for use. An Orby terminal waiting to send a message from the Earth would take a channel assignment and send the data up. The satellite would then process the data and send it down to a gateway Earth station, which would use land lines to feed it to a network control center in northern Virginia. Then the message would be rerouted through phone lines or other public switched networks to whoever the message was intended for or back to an Earth station, linked to a satellite again, and so forth until the message found its destination.

One area posed the most trouble because it couldn't be easily characterized: the accessibility of channels for the uplink. The question was, were there enough open channels at any moment for Orbcomm to pass messages without interference or delays?

Squiggly black hairlines on King's printout gave the first glimpse of radio interference that Orbcomm-X had promised to map more than fifteen months before. But instead of illuminating full plots of the Earth, as a satellite orbiting 500 miles above the Earth would, the picture appeared fractured into smaller segments. Although the SR-71 had flown 80,000 feet above the continental United States, it was not high enough and did not fly far enough to catch signals along the East Coast, nor did it stay in the air long enough to capture anything but a snapshot of spectrum use during one particular moment on one particular day. It was as if King had lofted an aeronautical flashlight into the sky and recorded everything the light illuminated on Earth as it rounded its circular path from Lancaster, California, east to southern Idaho into Canada, and back again. It was a snapshot, nothing more.

"See," he said, showing Steffy the graphs, "it starts here in California and went clockwise. So we get L.A., Des Moines, Chicago . . . "

Steffy flipped through more graphs. "But we still don't know about the rest of the world," he said.

Outside the United States, where foreign governments often failed to enforce radio frequency regulations, the picture probably would appear much more chaotic, reflecting a clutter of secret government communication systems and illegal interferers. That was the rub, another uncontrollable problem that added to Alan Parker's suspicion that the world was full of snakes.

In fact, it was widely rumored that hundreds—perhaps thousands—of independent taxicab companies and drug runners around the globe had settled around 148–150 MHz for unlicensed radio communications. Although no evidence of them appeared in the portrait King held, and certainly no unclassified data existed to prove otherwise, an accumulation of undocumented interference from spectrum squatters in the rest of the world might degrade the Orbcomm system from the start and prevent effective satellite message-passing in the band worldwide. At the very least, suspicions were reasonable enough to validate Parker's demand that King build two more satellites and test the bands in orbit. Any potential investor would want the same assurance.

Steffy looked up and shrugged. "The good news is—?"

"The good news is, there are still lots of other empty spots," King said. "Probability that the channels will be occupied constantly in this part of the world is pretty darn low. All we need are six free spaces. And looking at the data—see, right here are three free channels. And over here . . ."

"And it's mostly government use?"

"Logistics support, a little tactical communications, but not strategic defense. Mostly military police around bases talking to each other, actually."

Steffy passed through the pages again.

"So were Orbcomm folks happy with this?" he asked.

"I can't tell yet," King said. "Probably saying it's insufficient data."

"Probably."

"But it's better than a kick in the head. I mean, this band could be worse by factors of four or five and still be okay."

"You don't have to convince me. I'm hoping we could use these plots to show Orbcomm that we can ease our specs."

"Well, apparently it's not good enough to keep from demanding another Orbcomm-X mission. Still, as a first cut, they've got to admit the use in this band is light."

But in fact Parker and the staff at Orbcomm Inc. would not admit anything of the sort. Where King saw uncluttered airwaves rising over the Earth's horizons, Parker and his team saw ugly portents and evil omens.

King continued to complain about Parker and Orbcomm Inc. for a while, but Steffy never offered a word in response. On that subject, at least, he had learned to play it cool.

The next afternoon Steffy stood in Boulder's warm parking lot peering up at a five-foot section of white PVC pipe strapped atop a ten-foot yellow stepladder. A couple of engineers balanced and cross-braced the latest antenna model, which stood vertically as high as the pipe when it was balanced atop two sawhorses. The antenna looked like an artificial Christmas tree shorn of fir and tinsel, a thin sapling with only eight naked branches reaching out awkwardly from a narrow graphite trunk.

One of the engineers used his teeth to rip off sections from a grapey strip of homemade fruit leather and offered it around.

"No preservatives," he said to Steffy.

Another engineer dressed in Levi's and a purple cotton T-shirt huddled behind a tall stack of spectrum analyzers and RF gear, squinting into the green monitors, moving his hands to shade the glare of intense autumn sunlight. King's eleven-year-old daughter danced on the front steps, while her dad also came outside to examine the setup.

His engineers had emptied the parking lot of vehicles to limit the effect of radio waves ricocheting off their trucks and vans. Steffy circled around the yellow stepladder and pondered the test. King's team had created the kind of monument to cut-and-try engineering that immediately absorbs, as Steffy would say, "the interest of the student."

He spent a few minutes eyeing the kludge, but he could not fully comprehend it. Radio signals and fine points of antenna functions might as well have been a sorcerer's brew to him—satellite alchemy, for sure.

"So why can't we do this test in software?" Steffy asked.

King looked at him dubiously.

In Virginia, Steffy's engineers used software to launch the first Orbcomms for an entire week up and down the hallway. Draftsmanlike portraits of Orbcomm eight-stacks appeared on personal computers quivering under the force of a mock Pegasus launch. Freshouts would gather in doorways to watch a graphics simulation transform the rocket's nose into an animated, translucent vessel, giving a clear cartoonlike display of every physical deflection, twist, and shudder of a stack of eight Orbcomm satellites. Examining torsional effects for each millisecond of their simulated launches, the young engineers in Virginia noted structural pieces that needed strengthening and debated over moments when the satellite looked like the interior view of a queasy stomach jostled by stress. Using their opaque models, they tracked fluctuations of as little as 0.15 of an inch. The most dramatic twitch of red cylinders, blue brackets, yellow aluminum honeycomb—trembling in 3-D detail—reflected on their faces.

Here, by contrast, the first antenna test stood prominently over the parking lot like a totem pole for a society of ham radio hackers while squirrels played in the parking lot and popped red crabapples loose from the trees. Compared with sophisticated software tests at Sullyfield, Boulder's flimsy hardware test setup looked like an awkward architecture of Tinker Toys.

"We've got software for testing antennas," King said. "It only depends on how much faith you've got in it."

"Then why are we out here in the parking lot?" Steffy asked.

King's answer, in a look, was definite: No faith.

While autumn leaves rustled across the pavement, Steffy helped lift the antenna and move the sawhorses ten feet north of the stepladder. A probe on top of the ladder would measure the strength of the waves whenever an engineer at the power supply excited the top elements on the antenna.

"Rotate it," said the engineer squatting behind the monitors.

Steffy clocked the antenna 45 degrees.

"No good," the monitoring engineer said. "Try drooping the elements."

Steffy couldn't reach high enough to tug at the "branches" on the tree. "What if I clock it another 45 degrees?" he said.

"Yeah, but drooping the elements will help us get more power into the side lobes."

The guy with the fruit leather came over and handed Steffy a few pieces of paper, pictures of antenna patterns. He explained what the project manager couldn't see, critical shapes of the antenna's directional energy, as it pumped out radio waves in three dimensions. The 3-D drawings looked like an artist's conception of two zeppelins converging simultaneously against a white sky. Two strange lobes, drawn in a meshed pattern background, demonstrated the shape in which RF energy poured from the antenna.

"This is the pattern," he said simply.

Steffy pretended to vomit. Then he motioned to King.

"A parking lot doesn't exactly make the most wonderful test range," he complained. "I mean, here's the building, here's the ground, there are the trees, not to mention somebody who forgot to move his Range Rover over there."

King took the papers out of Steffy's hands and explained calmly: "For what it's worth, when you drive this thing with 90-degree phase, it's a circular pattern, but then you try to switch to a sleeve dipole at the end of the solar array . . ."

"Oooh," Steffy moaned. "Let me settle my stomach here."

But King kept it up, explaining that the test was only an effort to tune the four smaller reflecting elements, two pairs of "branches," at the bottom of the Christmas tree. Once they moved the elements into proper accord, his team would find ways to reflect the energy in a pattern that would yield two symmetrical lobes, closer to the ideal of having each pointing at 62 degrees off the horizontal axis. At the very least they could then build a scale model that could be tested in one of the anechoic chambers in the labs at Ball Aerospace, a few miles away.

King slung a tangy hash of RF lingo—cantation angles, circularity, axial ratios, polarization theory, azimuthal variations, and multi-path effects—concepts that made Steffy's mind reel. He felt himself being led quickly out of his depth, and he wondered if he knew enough technically to notice when he heard excuses.

"Look, Jan, here's the bottom line," Steffy said finally. "You have to appreciate that this topic is finally getting a fair amount of lofty attention now. We got everybody pumped up about antenna testing, and for the last couple of months all we've heard is this 'oops' plopping sound coming out of here. This might be normal playing in the clay, but we've got people running around asking us all the time, 'Tell me about your antenna problem.' I want to make it sound at least like we have some kind of coherent plan. I mean, say 'We're going out on the range at Ball on Friday,' or 'We're going to droop the Christmas tree.' That's a plan. But to say, 'Hey, we're going out in the parking lot and kind of futz around,' that's not exactly the kind of thing that warms the cockles of a division manager's heart."

The secretary appeared on the doorstep and called. Another phone call from Bob Lovell.

Steffy stayed with the engineers for an hour, fiddling with the equipment and carting the antenna around, collecting data from points near and far. He hadn't taken so much time to play with satellite hardware in months. Finally, when it seemed clear they would bog down for half the afternoon, Steffy decided he had better find King.

In all likelihood the antenna presented them with more than the usual engineering dilemmas. It had all the makings of a science project.

The light changed from the unusually bright October orange outside to a gray tint as he walked through the lab toward the main office. He felt comfortable barging in with point-blank questions. It was his duty, in fact.

The earlier message had been from Virginia, suggesting that the SR-71 data was, for all practical purposes, politically worthless. Instead of easing minds, the tests actually had persuaded Alan Parker to press ahead for tighter specifications on the satellite's communications system, an addendum to their impending contract that would, in all likelihood, force the Boulder crew to redesign the satellite's entire communications scheme.

For a moment Steffy was strangely silent.

"So the question," Steffy said at last, "is what do we want to sign up to as the spec?"

"You've got to be kidding—I want out of this spec completely," King said.

"I'm sorry, they won't let us."

"But look at the pad they want! I've never seen such conservative numbers!"

King insisted he knew something more about interferers in the band than Parker, and implied that he knew more than anyone outside the spook culture. The truth was, he actually had launched a number of Oscar satellites that used slices of band around 139 and 149 MHz for communications. Interpreting data gathered by him and amateur satellite users around the world indicated few problems. The most nettlesome interferers for the Orbcomm system would be amateur radio users themselves, he guessed, around 144 MHz. Federal government interference would crop up around 151, then military push-to-talk and aviation radio channels just above that. But nothing so severe that it required the unforgivably strict specifications Parker demanded.

Steffy knew what Alan Parker thought about King's AMSAT experience. He listened politely, then rephrased the question: "So how do we rewrite Orbcomm's specifications for the contract?"

"The way I look at it," King said, "we don't accept these numbers as a specification at all. Consider it a statement about good design practice."

Steffy groaned. "Come on, Jan, this a contract. We might think we can interpret these numbers any way we like, but our lawyers won't. How do we set the spec?"

"You seem worried." King smirked.

"I am worried! And I'll be even more worried when we need to start integrating parts and software and we still haven't agreed on the spec."

"The first units are supposed to be built by November 15—I'd say that's probably the appropriate time to get worried."

Two weeks! Steffy could ignore the remark, but he couldn't disengage. He looked out the window and stared silently.

"Be reasonable," King said, more earnestly this time. "Equalize pain among disciplines. That's always a good solution. You got pain, I got pain, Alan Parker's got pain."

"Just try to get the friggin' thing to work," Steffy said tartly.

"Okay," King countered. "But by the way, we only have about ten other things in this system that can screw us up. Or have you forgotten? We have the subscriber transmitter, the subscriber receiver, the gateway transmitter, the gateway receiver, adjacent spacecraft like Tiros, not to mention adjacent transmitters like Guatemalan taxicabs and drug runners for the cartels in South America—everything that's not part of our service. There's potential for interference in at least 40 percent of that. Raising the spec only increases the risk everywhere else. You tell me what there's *not* to worry about."

Steffy came back quickly: "And I'll tell you what the customer believes. He believes he's given us a conservative specification. But because of Orbcomm-X, nobody really knows."

Orbcomm-X*:* Jan King's brow suddenly crinkled. How could he sign up to specifications that would cause his team to fail again?

After a long pause he gave up: "Well, if we can't change the laws of physics, I suppose we better collect more data. They'll have their satellites by Thanksgiving. . . . I will personally bolt them on for the next Pegasus launch."

King's eyes held steady on Steffy's. He was making a promise, but he also needed an ally.

"Believe me," he said. "if those guys want another mission where the satellite doesn't transmit any information to the ground, they can count me out. I'm out! Our guys have already had one mission like that—I live in mortal fear of this—and I've already had my worst fears realized. I don't want to do see it happen again."

Steffy detected the fear. The insistent press of Alan Parker's demands, he thought, stemmed at least in part from professional jealousies. Danger of that kind of conflict intruding on his program was potentially greater than he wanted to imagine.

King turned back to his phone to call Lovell with news of his decision.

Steffy headed down a corridor to meet with other engineers in a wood-paneled conference room, where he picked up the name of a talented young guidance specialist from Northrop who was eager to leave his job. Mike Dobbles's incessant harping had finally encouraged Steffy to find someone experienced enough to take over the satellite's troubled attitude-control system.

At midevening Steffy raced to catch the red-eye back home. During the flight Steffy thought he would add a few more items to his "To Do" list, but then he set the work aside.

He cracked open a new book he had picked up about battles and military strategies of the Civil War and read intently. His own tender experience with combat and tactical maneuvers was at last underway.

CHAPTER SIX

1981–1992

A Little Walkin' Around Money

"Hey, man, check out our new flight computer," Denton said one morning, motioning to Gurney. "Pretty sexy, huh?"

Grace Chang looked up from her keyboard and shouted across the room, "So it's true!"

Gurney poked Denton.

"Wha—" Denton said.

"What I've always said is true!" Grace said.

"What's true, Grace?"

"Only men describe hardware as sexy."

"Huh?"

"Well?" Grace said.

She squinched her face mockingly, stared him down, and laughed before turning back to a technical discussion with a contractor on the phone.

Denton leaned over the workbench and whispered, "What's with her?"

But Gurney was having his own problems.

"C'mon, man, you've been playing with that board all week . . ."

Just then the project manager walked in and complained that the lab looked like a pigsty, much worse than labs used by the company's other satellite teams.

Denton objected: "I can't stand it when you compare us to other programs, Dave. We have two technicians living with us full time and both of them are slobs, and . . . Besides, it would be much nicer when you criticize us to substitute 'pig sty' for something less offensive."

"I can appreciate that," Steffy said. "How about 'porcine domicile'?"

"You know, we practically live here day and night."

"But you're still more than twice as messy as the other programs."

"Why don't you make that an exponential function?" Denton said jeeringly.

"Okay. You're four times as bad," Steffy said.

"I was just joking, Dave."

"No," Steffy said, grinning now. "C'mon, want to go for logs? Maybe we can make it cubic?"

They had not yet passed that important point separating the initial months of chaos and a midlife of steady efficiency. Drafting papers swamped the lab. Squeezed behind a tall mound of signal generators and oscilloscopes, Gurney watched his partner burn in one thin red flywire after another into his electronics. Every subsystem in the lab showed a similar tangle of red wires looping over the surface of their first boards, evidence of misconnected circuits, on-the-fly redesigns, or poorly manufactured products from vendors. Rob Denton's board, for instance, had come back from the manufacturer with transistors set in backward and with one of four layers in the board so severely skewed that he had to reforge connections using two hundred flywires and a soldering gun.

It sometimes made me wonder, despite David Thompson's boasts about hiring the great Next Generation of wolf pup engineers, whether their project was really that much different from other aerospace projects since the 1950s. In some ways, I concluded, they were not so different, given historical accounts I had read going back as far as the Vanguard project. At Orbital engineers not only shared the same quest for the American dream and the thrill of building a revolutionary system from scratch, but as was becoming increasingly apparent, they also had many of the same problems—schedule pressures, straggling vendors, too few hands, not enough experience.

A recent report on engineering accountability by the air force put the team's struggles in context. Throughout the business, the report said, the root cause of problems was that spacecraft were designed near the ultimate level of performance for given levels of cost and schedule. Excessive optimism prevented engineers from exploring designs with slightly lower levels of performance that could lead to lower program risk and lower costs and shorter schedules. Sometimes design feasibility wasn't even known until late in the development phase, when trade-offs had to be made. By the time flaws were exposed in testing, it was sometimes hard, if not impossible, to alter the design.

They were simply being engineers, like engineers of any age, in all the word conferred: hatred for sloppiness of detail—not so much in language or dress, which was an endearing eccentricity among many, but in the entrails of software code or in the smallest assumptions of a kludgey idea or hidden on the back of a motherboard where someone did a sloppy soldering job. They could criticize such messiness endlessly, and they did. They were the world's harshest critics, impossible to impress, a heady mass of human computational power and some of society's most conforming nonconformists. They knew that even if a system reeks, you still can't change the laws of physics, and in learning to respect such truths, they had earned a strangely philosophical bent that shaped a work culture that was both absurdly delightful, even childlike, and annoyingly hard-driven and cynical.

Madras shirts and ho-daddy haircuts also made me look twice at Steffy's stable of recruits, young engineers who not only communicated in a swirl of acronyms ("Your LED for the CPU on the ACE EDU was DOA after the BEEM, okay?") but also spent their occasional off hours very deliberately making babies and assessing home mortgage options ("We optioned the five-year ARM at three percent over a ten-year balloon and the fifteen-year jumbo at six percent, then dropped it for a thirty-year fixed with a three-quarter cap, but then Julie met a guy at DSI who got a ten/one product with a one-eighth cap and a forty-five-day lock at five-and-a-quarter, so we're haggling with the bank.") They did everything with the same mindbending joy and alacrity as their predecessors in the 1950s, when everybody, it seemed, was pregnant. Sometimes I really had to keep looking at my watch to remember what year it was.

They pumped iron at Gold's Gym, ate Chinese at Mom's across the street, drove every kind of hip machine from spunky new Firebirds to broken-down Saabs, wore the glossiest thin-soled wingtips and stylish leather, suede, or shiny synthetic jackets. Although nothing they did reflected the sort of campy interest in science fiction and atomic power or the dreamy affection for spaceflight that one might have expected, Steffy's troops did exude all the spry optimism and distinct joie de vivre about science and engineering that I imagined the first generations of space cadets did.

However, the question about whether the Orbital Sciences Corporation really was or could be different from any other aerospace company remained pertinent, perhaps critical, in their minds. In one way or another, it continued to come up in conversations every day around the labs where the company's Freshouts—and veterans—worked.

Given the sharp political currents that had shaped space technology since Sputnik and the momentous twists in cold war history since Apollo, it was generally assumed that the nature of global politics in the 1990s had left a definite division in the satellite business as it was practiced in the 1980s and as it was practiced in the last decade of the twentieth century.

The truth, of course, is that the history of engineering is rife with ironies, absurdities, serendipities, and parallelisms, which is, to a great extent, what makes it interesting, and that engineering springs from practical concerns—though often idealized—making even the most advanced technology as much a reflection of its time as a novel or a painting or an illuminated manuscript. Sociologists and historians have often pointed out that new technologies are socially constructed, shaped by a dizzying variety of forces. The influence of politics, economic interests, history, and competing technologies has far more of an effect on the shape of any instrument or any project, no matter how "new," than its creators might imagine.

Rooted in history and traditions, engineers also often find that they must depend on the rediscovery of common materials and the revival of old ideas. That was certainly true at Orbital. Low-Earth orbits, in fact, had been used years before the rush to geosynchronous orbit. And the whole aerospace business had begun with small satellites, after all. Even David Thompson would say he expected nothing really extraordinary from the first Orbcomm satellites—their very simplicity as data-only systems made them not so remarkable. He called them, in fact, "bongo drum" technology.

Still, it would not have been safe to assume that Orbital was just another aerospace company or, perhaps, that it ever would be.

Strolling through the company's high bay, chatting with the young engineers, there were times when I thought the project looked physically little different from what an observer would have seen inside any satellite project twenty years before. The clean rooms, white bunny suits, dessicants, tanks of hydrogen, strain gauges, wrenches, screwdrivers, hammers—nothing so revolutionary there. They still could not completely test an attitude-control system on the ground, nor deploy antennas and solar arrays in zero gravity. Even the engineers' motivations, their goals, the way they thought about space—all seemed in concert with their counterparts who had come before.

But the belief that they were in the beginnings of a major transformation captured everyone's imagination. Orbital's executives not only expected to pioneer the commercialization of space during the 1990s, but they also attracted young engineer idealists with expectations that were as unbounded as a space probe. Thirty-five years after Sputnik, entering the lives of a few dozen young Freshouts at Orbital's labs and offices, it was common to hear engineers laugh and complain about the "old" culture of aerospace. Taking advantage of a remarkable prevalence of exotic composite materials, computer-aided designs and simulations, and a vast array of miniaturized electronics, Freshouts boasted that they would "do" space right for the first time. They demanded quicker turnaround times from their vendors, insisted on truly competitive bids, and literally searched the world for smaller, lighter, smarter materials. They believed they were building the first personal satellites—like the first Apple computers—for the next milenium.

Certainly it was not too soon to question Orbital's potential to break radically from the past. At least one way—as entrepreneurs—the company's executives already had.

Something else, too: The Orbcomm satellite itself turned out to be, if not a profoundly new creation, at least a very difficult machine to build, far more of an on-the-edge design than David Thompson imagined. Every day the challenge left its engineers gasping for an ounce here, a milliwatt there, a few more centimeters of floor space inside the body of the spacecraft in which to squeeze their already tiny component parts.

All in all, difficulties only excited the team, encouraging Steffy's young engineers to believe they did stand at the avant-garde, on the verge of a revolution, in an entirely new place under the sun.

And like all idealists—mirroring the company's first small team of revolutionaries—they pledged to work like hell to make it so.

The original dreamers opened the Orbital Sciences Corporation for business in 1981 in a two-bedroom townhouse in Thousand Oaks, California, a suburb of Los Angeles.

Early in the morning when the phone jangled with calls from the East Coast, David Thompson would nudge his wife out of bed.

Stumbling to the phone, she glowered at him but answered sotto voce, like a cheerful secretary: "Good morning, Orbital Sciences Corporation. How may I direct your call?"

On those days "Mr. Thompson" was still curled up under the covers at six o'clock. Somehow he managed to parlay business with nothing more than a phone line, a ream of quality letterhead, and three maxed-out credit card accounts, two of which belonged to friends from Harvard Business School, Scott Webster and Bruce Ferguson.

The three partners did not even live in the same state, but held day jobs in different regions of the country—Ferguson with a Chicago law firm, Webster with a high-tech engineering company in Seattle. Thompson had a prestigious job as special assistant to the president of Hughes Aircraft Company's Missile Systems Group in L.A.

But for each young man the real goal was entrepreneurial, and together they conspired in an unlikely scheme.

Since 1977, when he left the California Institute of Technology, Thompson had taken jobs where he saw for himself how traditional aerospace functioned, first as project manager on the space shuttle's main propulsion system, then as an engineer on advanced rocket engines at NASA's Marshall Space Flight Center in Huntsville. At the time most large companies concentrated their efforts on defense systems and commercial aircraft, the most lucrative aspects of the aerospace business. Few concerned themselves with the less profitable—and much riskier—matter of creating products for commercial space. At the time NASA also was not under great pressure to lower its costs or to create technology for down-to-earth uses.

In fact, Thompson recalled later at Harvard Business School: "What I found in working for NASA was an enormous amount of technology developed by the government for a small number of programs. The technology worked okay, but that was about it, and none of it was being used in many areas where it might be. NASA was interested in its traditional mission, which was to explore space . . . and they really didn't have the sort of approach that would lead them to drive down costs."

Because of a lamentable tendency toward perfectibility and a definite aversion to risks, the most up-to-date technologies entered the aerospace industry so slowly that potential benefits from the explosion of innovations in the computer industry rarely found their way into "new" products. A small, private commercial company, Thompson theorized, could take advantage of technical trends far more effectively than any government agency or existing aerospace giant.

Knowing next to nothing about business, he entered Harvard Business School in 1979 to test his theories and think about his future. There he met Bruce Ferguson and Scott Webster, section mates who shared a common interest in space development. Together they worked on a NASA-sponsored field study assessing commercial ventures, particularly in the area of materials processing in the microgravity environment of space.

Their study eventually concluded that small companies were not particularly well suited to be prime movers in the industry. For one thing, any start-up venture would have to depend heavily on the support of NASA, an institutional relationship that seemed unlikely for any but the largest companies. Most importantly, the start-up costs were astronomical. In order to prove themselves capable, leaders of any small aerospace company would have to consolidate complex deals by forging partnerships with large investors. It would be particularly important to woo significant financiers and influential industrialists who could help the company cross the enormous capital barriers (for example, $50 million for a launch). Such a goal seemed implausible for any new business. Even in the initial stages of a commercial project, they concluded, the risk was so high and the payoff so uncertain that few investors would sink money into entrepreneurial ideas.

Surprisingly, Thompson, Ferguson, and Webster were not discouraged by the outcome of their work. Quite the contrary, they believed the research left them with a unique document containing an exceptionally clear set of observations identifying every impediment to a commercial space business. If the time was not right for materials processing in space, they expected that someday conditions would improve. Identifying risks and describing the pitfalls of a commercial business in space, they had, in fact, created the outline of a vision. They had, quite ironically, predicted failure and simultaneously laid the groundwork for new opportunities.

Six months after graduation, David Thompson took a call at work at Hughes Aircraft's Missile Systems Group. The Space Foundation had awarded their school paper a first-place prize for contributions to the understanding of commercial space activities. Unknown to the three friends, their ideas were quickly circulating within the larger aerospace community.

When they met in Houston soon afterward to accept the award at the exclusive River Oaks Country Club, they found themselves surrounded by a crowd of interested, generous, and wealthy people, a combination of Houston's upper crust and several of the country's headliners from the old Space Age.

"We were all having a good time, walking around this swanky place, with an oilman on one side and Alan Shepard on the other," Thompson later recalled, "when the conversation drifted to future prospects." One of the oilmen, a Texan named Fred Alcorn, told the "boys," as he called them, to stay in touch. If they thought of a serviceable idea, Alcorn said, he might just help them out with "a little walkin' around money."

"A little walkin' around money?" Thompson wondered. "What's that?"

Writing a paper for school and starting an aerospace company proved to be distinctly different problems. Even with the prize and the unexpected interest of enthusiastic boosters, the three young men encountered sudden barriers whenever they met with aerospace officials. Most took one look at the "boys" and walked away. The corporation itself resembled an *Our Gang* fantasy: young innocents planning a moon launch from a sandbox. Anyone with a view of what actually went on behind the corporate facade might have had the same thought—that the only difference between Thompson and a boy in a sandbox was that at twenty-eight, Thompson should have known better.

Although he occasionally managed to find his way into a presentation with NASA, once there he usually met with tremendous skepticism. In fact, one high-ranking government official at NASA later admitted that after seeing David Thompson the first time, he thought: "My gosh! This guy doesn't have a pot to piss in or a window to toss it out of!"

Aerospace executives viewed start-ups with amused detachment, at best, and for good reasons. For example, if a satellite was nothing more than a highly placed repeating antenna, the cost of making that placement presented a most devilishly clever deal. As simple as it might be to raise a 100-watt antenna on Earth—little more than a run to the corner Radio Shack—planting an antenna in orbit generally required a $50 million rocket ride, oversight from a half-dozen federal agencies, a standing army of engineers, technicians, managers, and several well-placed lobbyists and lawyers in Washington. Space ventures made fine sport for sovereign nations and corporate behemoths, but the rules of the game usually prevented would-be mom 'n' pop outfits like Orbital from taking the first step onto the playing field. The problems sounded familiar—straight out of their Harvard field study.

Interestingly, though, while Thompson and his pals continued rummaging for the right idea, the political climate in the country changed. If only briefly, incentives created under President Ronald Reagan's administration in the early 1980s allowed a slim chance for new businesses in aerospace to succeed. Specifically, the Republican administration in Washington led a battle to create tax incentives for investors in commercial space projects.

Although the shuttle was selected to be the nation's exclusive long-haul trucker to space and there was no call for anyone to develop an expendable launch vehicle, the shuttle did have a number of specific limitations, which Thompson had recognized during his experience as a project manager on the vehicle. Those limitations, he suspected, could create opportunities for a few small companies to step in with products that would enhance the shuttle's effectiveness. With a handful of tax incentives to give a boot up, Thompson expected that if he and his friends could find the right product to sell NASA, investors would come calling.

While imaginative possibilities for new products seemed endless, the real challenge was predicting trends. They felt sure they could invent accessories to support the spacecraft, but the toughest job would be to convince leaders in government, banking institutions, and a few aerospace corporations that they should serve as partners in an essentially start-up enterprise.

The idea they settled on came from a space technology conference in Washington in 1982. Still following the blueprint they produced in business school, Thompson began visiting NASA headquarters frequently, and during this one particular conference, several people predicted that because the shuttle only flew in low-Earth orbit (400 to 700 kilometers), there would soon be more missions when the shuttle would require a new kind of machine that could launch satellite payloads into higher orbits. Essentially, NASA needed a booster.

Also, at the same time, Thompson discovered that a number of companies had already planned new lines of communications satellites to be launched from the shuttle. They, too, would require a booster. Targeting that class of satellites, David Thompson and his friends decided their first product—their first entrepreneurial offer at the offices of NASA—would be a machine they called Transfer Orbit Stage (TOS), a relatively inexpensive, upper-stage booster for large spacecraft.

Anyone who sat down to do a brief financial analysis would have concluded the young men would never make it. The reason: Huge capital barriers to entry stopped all but the wealthiest companies from bidding contracts; the most powerful customer base in the world, the United States government, had created an exquisite financial straitjacket in the form of risk-defining rules that led to expensive, redundant systems and large capital outlays for, of all things, paperwork; finally, some of the most successful technology companies in the world would necessarily be either Orbital's suppliers or its competitors. Given these conditions, venture capitalists would have seen little to recommend Orbital.

Once again in 1982 a political change gave them encouragement. From the White House, as President Reagan began to tout commercial ventures in space, NASA administrators took the president's rhetoric to mean they should demonstrate an increased interest in commercial projects. Combined with new federal tax laws that allowed investors to take greater risks with high-tech start-ups, little Orbital Sciences Corporation, comprising three unpaid "employees," stood a chance of gaining a toehold.

For instance, tax laws had changed so significantly that a financially sophisticated investor with, say, a $200,000 annual income or a million-dollar net worth could easily write off losses on a high-risk start-up. A $50,000 investment allowed a $27,000 tax deduction. For high rollers, it was a good gamble, like investing one dollar for two or spending fifty cents to get a dollar's value. The leverage on investments was potentially so rewarding that Orbital began to look like an attractive prospect.

Whereas to the untutored ear, the term *Commercial Space* sounded like an oxymoron and *entrepreneurs in space* simply sounded moronic, the two concepts suddenly gained the jingle of hard currency. As fortune would have it, the wealthy Houston oilman Fred Alcorn came through like an angel. He made good on his promise by stepping forward out of a flush season in the drilling fields and gave Thompson, Ferguson, and Webster (still "the boys") $300,000. Because Orbital had only the three employees at the time (not counting an apartment neighbor in Los Angeles whom Thompson was now paying a few dollars to answer his phone) and $1,500 in credit card floats, the oilman's investment came like a rain of riches. For the first time, they raised their profile among investors. With Alcorn's small investment and no lack of chutzpah, Orbital's founders boasted that they, as a private commercial outfit, would build a satellite booster faster and cheaper than anyone else in the business.

First obstacle: They not only had to convince NASA of their potential, but they also had to assure a number of other institutions that their little business—touting itself as a "technically based management, marketing, and financial corporation"—could handle the affair. The founders called it their "simultaneity problem." That is, they had to persuade NASA to let them launch an untried booster (a bomb, essentially) inside its billion-dollar space shuttle. They also had to prove that its controlled explosion would not detonate before scheduled. They had to form a partnership with professional satellite builders who would allow them to attach this "bomb" to a semiprecious, multimillion-dollar spacecraft, and at the same time sign contracts with hardware suppliers who were willing to invest in research and development for a brand-new product. Most significantly, they had to raise $50 million.

Every academic, financial, governmental, or corporate institution they called on wanted to see three or four others signed on first. Forging alliances with conservative aerospace companies and finding champions among risk-averse government bureaucrats was the only way they could possibly make the project appear consistently attractive to investors. It remained a preposterous goal for three boys just out of business school.

Unlike start-ups in the computer business at the time (the story of Apple Computer became a common comparison, in their minds), the Orbital founders could not visit their neighborhood electronics store and say, "Hey, we can build this neat little product, and if you'll take a hundred of them on consignment, we'll split the profits with you." Instead, they had to go directly to NASA and directly to the U.S. Air Force and say, "We have an idea for a great new product, and it will only cost you $25 million, and we know if it screws up the first time and wastes one of your satellites, it's another $50 million loss, but you can trust us because, well, we're smart and we have good intentions."

In fact, it did not matter how good the idea, how worthy the engineers behind it, or whether the Harvard-educated entrepreneurs came to them as prophets from God or creatures from another planet, the barriers still stood so high that their prospects remained fundamentally absurd.

Nonetheless, the head of NASA's space shuttle booster program, Jack Wild, finally agreed to meet Thompson and friends in late 1982 at an aerospace conference in D.C. Although the Orbital trio offered to provide their own commercial version of a booster at a much lower cost than traditional government-financed projects, Wild scoffed. Only because of President Reagan's newly stated interest in commercial ventures did he promise to introduce them at a meeting with his staff in December of that year.

"I figured I'd give these kids a quick hearing and then throw them out," Wild later recalled. "I had a pre-meeting [with NASA staff]. To a man, they were extremely negative—you know: 'Who are these guys?'—that kind of stuff."

Instead of being booted out, Thompson was allowed to present his ideas fully and spin his charms. Remarkably, it worked. "He blew their socks off," Wild said. "Each one of us did a 180-degree turn."

At that first meeting Thompson made such a convincing case and the three young men, as a whole, left such a strong impression that NASA gave them six weeks to prove they could handle the project—financially, technically, logistically. They signed a memorandum of understanding (MOU) that allowed NASA to monitor the development of TOS, but only if Orbital could support the venture itself with private rather than public funds.

The response to NASA's unusual arrangement was also remarkable. Alcorn, the Texas oilman, secured a $2 million line of credit with his bank. Executives at Martin Marietta, who had just experienced an unusual downturn in its business cycle, suddenly found themselves looking for a quick new product to make up for losses, and, seeing Orbital's surprising clout, as well as its remarkably shrewd new product, signed on as the prime contractor. The three "boys" pulled together a board of directors that drew on all of their highest-level contacts, including several former NASA officials and industry leaders who believed in the entrepreneurial promise of new space commerce.

At the next meeting with NASA, members of Orbital's newly created board of directors spoke with stunning confidence. Alcorn stood behind his line of credit and said there was more money where that came from; the president of the Space Foundation gave a stirring speech on free enterprise in America and its importance to the U.S. space program; and Martin Marietta representatives came through with a convincing technical pitch.

Though roundly impressed, NASA executives still responded cautiously. They did not give any assurance that they would ever buy one of Orbital's TOS boosters, and they warned that even if they did sign a contract with the company, they would continue to accept similar proposals from other companies.

Finally in April 1983, not long after another booster manufactured by Boeing misfired while trying to launch a satellite off the shuttle, NASA called on Orbital to sign a more formal agreement. If the little company could raise $50 million within five months, NASA stated, Orbital would get a contract.

In the short history of Orbital, every reversal of fortune, as it turned out, became a predictor of success. Often the founders' brashness alone brought favorable publicity.

For instance, when *The New York Times* wrote about the company's successful negotiations with NASA, the article caught the interest of Nathaniel Rothschild, a member of a well-regarded European banking family. Supporting an American aerospace venture fit in well with unrelated political maneuvers Rothschild had planned against the French government, so he lent Orbital his support, amounting to $1.8 million in venture capital, and helped the company begin its first heavy round of financial negotiations.

Bruce Ferguson, who had quickly become the company's financial wizard, at last had some leverage to raise $50 million for a little start-up company with no track record. A road show began with the three-man Orbital team making a tour of the country to meet personally with private investors and potential corporate partners.

"Wherever we go," Thompson confided to an interviewer at the time, "people look at us, and if they don't say it explicitly, their faces do: 'Aren't you guys a bit young to be doing this?' Well, you say what you need to say and move on. We mention Steve Jobs at Apple or Fred Smith at Federal Express. Or we say that commercial space transportation is such a new business that no one really has relevant experience. Or we tell them that because the project has such a long time horizon, chances are you'd die before the first launch if you waited for gray hairs before starting. Let them draw their own conclusions. My point is you've got to be bold, to make do with what you have better than the next guy. We've done that so far, and now we need $50 million in three months. I believe we're going to get it."

An extraordinarily complex set of negotiations followed. Orbital organized a research and development limited partnership called Orbital Research Partners, which offered significant tax advantages and up to a 500 percent return on investments. The R&D partnership, created with the help of Shearson/American Express investment bankers, allowed Orbital to sell $50 million worth of shares in a partnership and gave each investor a way to write off their investment immediately, with the hope that there might still be a profitable product created in the end.

The legal fees alone cost Orbital $800,000 in 1983, but by early 1984 the three founders had raised more than $50 million, the largest amount of private capital ever raised for an aerospace system at that time. It seemed possible that President Reagan's statement, reiterating his national space policy to engage entrepreneurs in the commerce of space, might hold some truth: "When the profit motive starts to play," the president had said in 1983, "hold onto your hats. The world is going to see what entrepreneurial genius is all about."

Less than three years later, in January 1986, the space shuttle *Challenger* exploded. The president's goal to field a new industry of space entrepreneurs dissolved.

The shuttle explosion resulted in a moratorium on all American space missions, and with the shuttle sidelined, a number of the promising aerospace businesses went belly up. Like many other smaller companies in the industry, Orbital faced life-threatening peril.

A week before the rollout of TOS during the summer of 1986, President Reagan also announced that NASA would no longer launch communications satellites from the shuttle. With that single statement, Orbital's business plan, which forecasted that half its revenues would come from launches of comsats, should have fallen to pieces.

As luck had it, just weeks before the Challenger disaster, NASA had ordered TOS boosters for the Mars Observer and ACTS satellite, mission contracts that left Orbital with a total $90 million in contracts. Those alone provided substantial progress payments to keep the business afloat for the next four years.

Thompson and his friends realized immediately, though, that they had to cast a net for another new product. And they had to do so quickly.

As a dark horse, it could be said, Orbital always ran very, very well. Its reputation expanded with each new and daring financial maneuver. The success of the first Pegasus rocket launch in 1990 brought multiple orders from the United States government. No one could believe that a small team of essentially amateur "space nuts," as *The Wall Street Journal* tagged them, could have created the first new space launch vehicle in twenty years. The vast publicity and abiding interest of NASA raised David Thompson's visibility even higher within the aerospace world.

Using funds from a combination of contracts for its TOS booster and a joint venture with Hercules Aerospace Company to develop the Pegasus rockets, in 1988 Orbital not only raised the private capital needed to develop its rocket but also acquired the Space Data Corporation in Phoenix, Arizona, a leading manufacturer of suborbital rockets, launch facilities, space payloads, and related data systems. Overnight Orbital became one of the most prominent launchers of suborbital rockets in the world, and it rose quickly from the ranks of small independent aerospace contractors to join the competition for launch services internationally.

The company expanded rapidly from thirty-five employees in 1988 to six hundred in 1990, to more than one thousand in 1991. By then David Thompson, like his business colleagues and all of his teammates on the Pegasus team, still thought it possible to transfer the elements of success from a particularly thrilling small-team culture to every level of the business.

They were still kids, but they had become, at least in the industry, legendary. The story of their meteoric rise was known by young engineers like Mike Dobbles, Steve Gurney, and Rob Denton from coast to coast.

CHAPTER SEVEN

February 1993
Popcorn Deals

Don Thoma dashed to the secretary's island, full tilt at the telephone. He didn't notice the company's two top executives standing in a doorway watching him run.

Under a shock of fluffy blond hair, his face looked tense. Cordovans skimming across the carpet, a plum-colored jacquard-weave tie looping over his shoulder like a coil of rope, Thoma spun around the corner and lunged for the receiver.

"Hold on a second," he said, calmly into the phone. Then, passing the receiver back to the secretary, he whispered eagerly, "Send it back to my office."

For the first few weeks in the job, he would talk to anyone—pilots, long-haul truckers, midwestern farmers, meter readers, scientists, oil well-head guys, ambulance drivers, merchant marines, car theft detectives, even a man who needed help with his first Toastmaster's speech in Kansas City. Anyone who dialed the new 800 number spoke to Thoma because he had an MBA from Harvard, and for at least a few weeks, he did more than work for Orbcomm's marketing department, he *was* the marketing department.

"Let's say you're the guy selling drinks on a resort beach in Acapulco and wireless credit card verification will not only make life easier but more profitable. . . ."

He heard his new colleagues working the phones as he hustled down the corridor. The 800 line attracted a three-sigma lot of loonies and curiosity seekers, but one out of five landed a legitimate query, tips that he was passing along to the new men in his office.

"I'll tell you what—go into your local outback shop, and talk to the guy who's selling pitons for hikers. Tell him about emergency messaging from the tundra, and watch his eyes light up."

The business known as Orbcomm Inc. occupied only about a dozen offices within earshot of David Thompson, whose own office anchored the end of the hallway. Intimate quarters invited eavesdropping.

"Okay, you say you're tracking stolen cars in Minneapolis? We can put you into a market where you can locate any car in the United States within 100 meters. . . ."

Thoma slipped into his office, tossed his coat across a gray armchair, and reached for the phone.

"Okay," he said, "yes, you're correct, it's fair to say our service in northern Canada will be less than in the U.S., at least initially, so if you're tracking animals . . ."

In the beginning conversations crackled with promise. If Orbcomm could not immediately provide continuous coverage in certain areas above, say, the Northwest Territories, Thoma would tell them. If someone needed technical information about software and radio links, he would give it to them. Sometimes he simply held on to the phone and started a list of modifications to customize the satellites for a particular job.

But he never dangled carrots. Alan Parker had been clear: The market should shape the technology and define the service, not vice versa.

Thoma kept scribbling on a yellow legal pad.

"Yeah, we'll put the first two satellites into polar orbit this year—that should give you the best coverage of northern latitudes. And by the end of next year, you can have intermittent communication with any place on earth. By 1995 we'll have a full constellation of thirty-four satellites, and you can go anywhere with it. And you say you're tracking what?"

Large animals in Manitoba.

He managed not to laugh out loud.

People wanted satellites for moving freight, delivering mail, tracking illicit drugs, and irrigating wheat fields. But this was something new.

While marketing managers at a half-dozen of the country's largest defense and electronics companies contemplated two-way, global wireless phone services created from large multibillion-dollar satellite constellations, they never saw discrete data markets springing up from a thousand different niches. Yet, Thoma was finding an explosion of businesses that only needed to send a short paragraph of information infrequently around the world, and of course no one wanted to pay more than a quarter per call.

"You did say 'large animals.' . . . And how many devices do you have in use out there right now? . . . No, I'm just trying to get a feel for your market. . . . Caribou? Okay. . . . And is that number 'population' or 'cost'? . . . You're paying $3 million right now? . . . And you already have caribou out there carrying receivers. But certainly . . . mmmm. . . . Why would a caribou need two-way communication?"

The swarm of calls energized him. One day he had General Motors on the line, the next day Ford. Westinghouse invited him to speak to executives in Maryland about a dozen applications. The U.S. Postal Service called for details. The Department of Defense wanted briefings. Farmers needed measurements of humidity and temperature in potato storage bins. The variety was astonishing.

Could a satellite track a couple hundred tractor-trailers freighting munitions, hazardous waste, *and* chickens up highways through the Appalachians? Could it communicate with Sherpa tribesmen in Nepal? Would it work on the Xingu River? How would it do with reindeer?

Until the last couple of months of 1992, Don Thoma was just one lonely guy in a dull office clutching a phone. The phone would ring in the morning and waylay him as the market began to speak. And what it said surprised the hell out of him: instead of calls for gee-whiz personal global communicators, people wanted a thousand different applications, quite different from what the executives at the big corporations had imagined.

Originally an engineer on the first Pegasus team in 1990, Thoma had left Orbital to enroll in Harvard Business School intending to mull over future prospects for satellite communications, for the Orbcomm business, and for its big competitors in particular. The chief competitor at the time, Motorola's LEO system, known as Iridium, appeared to be no immediate threat. Figuring its costs at more than $6 *billion,* compared with Orbcomm's relatively piddling initial budget of $135 million, Thoma concluded that Iridium's costs would be so extravagant that the service would reach only a small segment of high-end users—rich men on yachts, globe-trotting executives, celebrities with money running from spigots. Orbcomm's *everyman* standard—Thompson's "bongo drum" technology—would operate at least three to five years sooner, cost an order of magnitude less, and capture pedestrian markets that Iridium, for all its spark and glory, would choose to ignore.

That was one difference between the "big" LEOs like Iridium, and the "little" LEOs like Orbcomm. With an eye on markets for global telephony, the Iridiums of the world cared nothing about passing small bits of data from place to place. While the big LEOs slugged it out in front of the FCC for different frequencies and hungered for vast sums of money to put their systems into operation, little LEOs were left to exploit data markets for a relative pittance.

When he returned to northern Virginia with his newly minted MBA, wearing fine dark suits and spiffy cordovans, Thoma was thirty-one, enthusiastic and focused, eager to spread Orbcomm's tentacles across tundra, into deserts, around oceans, through distant mountain ranges. To his surprise, the market appeared more mysterious and elusive than he ever imagined.

After a few months his bookshelf looked like the industrial arts section of a local library, groaning under the weight of stony tomes about pipeline routes, oil rigs, and refrigerated railcars. One day I found him there, leaned back at his desk, engrossed in a book called *How to Succeed in Big-Time Trucking.* Instead of making its first big splash with an Orby personal communicator, the satellite system was sending taproots into huge, lucrative industrial markets comprised of hundreds of thousands of massive oil pipelines, heavy-duty field meters, lost cargo, stolen cars, and refrigerated trains. Blueprints for Orbcomm's skyland industry shifted from the *Star Trek* communicators to applications that would trace and monitor the circulation of almost anything—animal, vegetable, mineral—practically anywhere in the world.

Ideas for killer apps came from a most unusual assortment of clients—farmers, petroleum companies, army troops, truckers, environmentalists, Boy Scouts, biologists, airplane pilots—each of whom had an immediate use and many of whom wanted a customized application all their own.

Like him, the new marketing team spun through a weekly pattern of cold calls, sifted through a raft of messages from the 800 line, and flew to a mix of boat shows and telecommunications conferences from Montreal to L.A. They soon agreed that they had not identified vast numbers of people needing a personal global/mobile messaging device—at least not in the most populated areas of the United States. The cellular telephone industry, growing fivefold since 1985, was expected to capture mobile communications for every significant metropolitan area, and even though Orbcomm was not intended to be a metropolitan service, the stunning growth of cellular did pose a future problem if the industry realized there was a vast market for data communications—wireless e-mail—and discovered a way to complete its coverage in rural areas. Finally, the one market that Orbital fully expected to capture—emergency messaging—presented a risky quicksand of liability issues that no one had time to sort through immediately.

It was important, however, to see that Orbcomm had several natural advantages over cellular and paging services. Because cellular depended on the creation of small "cells" serviced by low-powered radio transmitters at every one-to-ten-mile radius, the industry covered only a small fraction of the total land area in United States. By comparison, Orbcomm's web of satellites would service virtually every spot of ground across the entire planet. Paging companies, even if they used satellites for passing data—beeping a particular pager and sending a phone number—had the problem of being one-way communication. Neither service could provide geographic location data, either, as Orbcomm proposed. Finally, as the market splintered into a bewildering array of wireless consumer services with a dizzy variety of roaming codes, pricing arrangements, and multiple phone numbers, the heart of the problem for any service remained how to offer reliable, low-cost, two-way communications anytime, anywhere. As a small data service initially, Orbcomm couldn't do everything perfectly, but it did have distinct advantages over existing wireless technologies.

By February 1993 a fractured market of industrial applications appeared likely to support the business most quickly, from the moment the first two satellites were launched. No one knew quite how to break the news to David Thompson and Bruce Ferguson that, for the first year at least, a mass market for Orbcomm's personal communicator—the real barn-burner—would have to wait. At the moment Orbcomm's barn-burner amounted to linking trucks and railcars and any kind of traveling cargo with their home bases or outfitting large farms with satellite communicators so a farmer could monitor the humidity of his potato bins or time his irrigation systems and check his sprinklers for leaks and breakdowns. It wasn't as sexy as saving lives, but as the first start-up services, they would bring profits with only the first two satellites in orbit.

When Thoma finished with the caribou caller, he hung up the phone and searched for Orbcomm's chief engineer, David Schoen. He had to compare his costs with the Argos satellite, already tracking reindeer in Manitoba.

"I thought at first the guy wanted to track fish," Thoma told Schoen. "I kept wondering how we'd fit a subscriber transmitter on a fish."

As they talked, Thoma suddenly noticed Bruce Ferguson taping another article to his door and window.

The chief operating officer regularly went to his door to paste news of camping accidents, mountaineers lost on remote mountain expeditions, and ordinary people freezing in their cars on isolated roads during winter. Ferguson, the company's cofounder, stood at his door and window taping tragic tales to inspire a humanitarian spirit, a sense of mission about the project. In the middle of the stories, he posted a sign in black marker: THE WORLD BEFORE ORBCOMM.

Too bad, given the satellites' potential for saving lives. But at the moment tracking caribou and monitoring potato bins made a better business.

Thoma didn't have the guts to tell Bruce that the world he imagined after Orbcomm would certainly be different, but not quite as the COO expected.

In the hallway someone stationed a full-size portrait of Captain James Kirk, commander of the starship *Enterprise*, holding an Orby. Underneath the captain's bold stance, it said, "Computer Log: Stardate February 3, 1993."

Thoma ran headlong into the display as he came glancing around a corner toward his boss's office. He looked quizzically at the cardboard standup and, beyond it, into a glass conference room, where two engineers stood struggling to send a message from a battery-operated Orby to a gray laptop about 15 feet across the room.

"I can't get the thing to work!" muttered one of the engineers.

This latest version of an Orby could only operate for 15 minutes under battery power, took an entire day to recharge, and frustrated the engineers endlessly.

As they cursed, Thoma spun around and slipped into Mark Dreher's office. His eyes lit on more images—sketches for the first ad campaign dangling from the far wall. Illustrations showed people pointing Orby antennas at the sky, with captions like: "It's good to know there's someone watching over you!" and "It's a direct link to the lumberyard or your office from anywhere!"

"What's the matter?" Dreher snorted. "Need a little more sizzle on that steak?"

His boss's brusque manner caught Thoma off guard.

"Maybe you're thinking more along the lines of—" Dreher gestured as if he were plastering new images on the wall "—some woman breaks down on the side of the road, the kids are crying, she's got grocery bags in the back, she's scared. God, just imagine making that pitch on TV!"

He spoke like a man who could sell army boots to the Benedictines—or boasted as if he already had. Loud and brassy, Dreher was not much older than Thoma, but he already had made his first killing out in Colorado by creating a large cellular radio service for the Rocky Mountain states.

As Orbcomm's first marketing director, Dreher sounded like an echo of his boss, Alan Parker—he did not care about satellites, rockets, or aerospace. Counting bytes as they passed through Orbcomm's network and collecting revenue was the only thing that would make him a happy man.

On another wall, directly behind him, right above photographs of his kid, hung a homely cross-stitched pattern in blue and pink: "Happiness is a positive cash flow."

"Well," he said, "how were things in the great Midwest!" Dreher plopped down in his executive chair and lifted a leg up on the desk.

For most of January Thoma had been absent, completing a road show from Texas to Idaho, visiting remote areas to pitch Orbcomm.

He had also joined presentations with the Federal Energy Regulatory Commission, interagency search-and-rescue planning meetings at the U.S. Coast Guard, and the Northwestern Policy Council in Idaho—not the most scintillating venues but good sources for tips about positioning the business.

Thoma had a uniquely persuasive style—he could burnish his delivery with rare intelligence—but he did not have the glad-handing personality of his more experienced colleagues. While Dreher, for instance, liked to tell people Orbcomm promised to be "slicker 'n snot on a doorknob," Thoma tempered his enthusiasm more gracefully, emphasizing the system's airtight engineering and homespun technologies. On this latest trip, he said, he had proved especially effective.

The week before, Thoma had toured glorious west Texas, stepping across dusty expanses of oil pipeline territory. As a young Easterner wrapped in a gray Chesterfield, creeping across remote and rocky fields—"centrally located," as he had been told by some jokester on the pipeline crew, who later explained that "centrally located" meant "400 miles from Amarillo, 400 miles from Albuquerque"—he amused his escorts. At one point, he told Dreher, one of the pipeline workers directed him over a series of barbed-wire fences that separated them from the pipes he needed to inspect.

"What are these footprints?" Thoma called back to the crew, as he crossed the last stretch of barbed wire. A coating of dirt dulled the sheen on his office shoes.

"Ki-yote!" one grizzled fellow yelled, and the oilmen piled into their pickups like they would abandon him, joking hysterically that they would leave him for the dogs.

"A real nice group of guys," Thoma said.

On the other hand, the oil company executives formed a more enthusiastic audience.

Best of all, the gas utility industry was being deregulated, creating a spot market for monitoring gas and pipeline capacity. Suddenly oil companies wanted to install flow meters on their wells to date production, regulate capacity and output, and begin to manage the lines more efficiently.

"They'll need data every day," Thoma said.

Dreher raised his eyebrows. "Every day?'"

New federal regulations had created the most immediate need with a requirement that companies collect data on the allocation of oil at any given well. If a company overproduced, it would be penalized financially. At the moment, a service like Orbcomm that could monitor flow meters on oil wells, date the production, and give the company a quick way to regulate the capacity or output on each line looked attractive. Also, in emergencies, an oil line fitted with an Orbcomm transceiver and sensors could uplink a message to a satellite, which in turn would downlink directly to the company or to an Earth station, which would route the message to the company, alerting someone of a leak or sudden falloff in flow. Thoma had heard of one company facing a $300 million lawsuit as a result of an undetected oil spill. Selling them a truckload of Orbys would be pretty easy, he figured.

"Look at it this way," Thoma said. "Even with just two satellites in operation, we can give them a hundred bytes every night and an alarm if a pump goes south or they spring a leak. If it's something like ten dollars a month for the service, that's attractive to these guys. They've got 40,000 wells out there right now, and they've got to have some coverage soon."

Dreher reached across his desk for a calculator and kept tapping at it while Thoma talked.

"That's half a million bucks a year," Dreher interrupted.

"What's half a million bucks?" Thoma asked.

"Ten bucks a month, that's half a million bucks a year! That's seven percent of our nut for 1995!"

"What are you talking about?"

"Wellheads!" Dreher said.

Immediately Thoma caught the excitement and said he knew of a company in Mexico preparing a bid on 20,000 wellheads as they spoke. "But naturally nobody's doing it with satellites," he said.

"The economics are wonderful!" Dreher cried, fingers flashing. "If we go out and make $25 on each Orby, and you've got, conservatively, 24,000 wellheads. Heck, that's $600,000 we never counted on! Tell you what, buddy, that should cover your salary, mine, and a piece of Alan Parker's."

"I thought we'd just be leasing the equipment to them," Thoma said.

"No way, if we can shift our market away from that high-value emergency messaging and turn our attention to these industrial-type popcorn deals, we can sell a zillion terminals. With minimal messaging. The revenue from the terminals alone leaves us a lot of gravy."

"And if we could actually poll the wellheads twice a day—" Thoma began.

"Or maybe we can do it like Gillette—give 'em the razor, sell 'em the blades," Dreher said, already steps ahead. "Wait till we start pricing! People will be knocking down the doors to tell us what they want."

LEO satellites would work perfectly for automated meter reading, they decided, where meters were remote and hard to reach. The cost to utility companies to hire people to read meters was said to be a significant issue. Misread meters, the cost of travel time, and environmental constraints made the job problematic.

Thoma also had discovered that buried oil and gas pipelines were subject to corrosion from electrical reactions with the ground. Cathodic rectifiers buried every couple of miles along the pipelines could detect minute electrical charges that signaled corrosion. At the time companies used the rectifiers to alert them for maintenance, but with more than 400,000 miles of pipeline—some running through harsh wilderness—it was difficult and expensive to send personnel out to check the rectifiers regularly. On the other hand, for the right price, a satellite might do the job quite nicely.

Dreher pushed his arms up at the ceiling and stretched.

"There are a hell of a lot of pipelines and utility meters around this country. Motorola chased that son of a bitch like you wouldn't believe, but it's been twenty years, and no one's come up with a good solution yet. Maybe we'll finally pull it in."

He paused, and the smile on his face just grew bigger.

"Thoma, my man," he said, "this isn't like selling rockets to the government. We're talking big revenues! I shit you not, this is looking real good!"

Mark Dreher loaded up his plate at the Pizza Hut buffet cart and stuffed a glob of melted cheese and tomato into his mouth even before we reached our table.

The pace of business reminded him of what it had been like ten years ago in 1983, when he and a couple of friends from Motorola started selling their own cellular phone service around Denver: Run "scattershot" for a while, then go "full bore" at major targets.

"I don't know a lot about satellites, and I don't want to," he said between bites. "Nobody cares about the technology. 'What have you done for me lately?' that's what I want to know. 'How can you make my business more profitable?' I mean, all right, the technology's neat, but so what?"

Also, like Parker, he suspected that the satellite engineers lacked the necessary sense of urgency, and the extra time they kept requesting sounded unreasonable. The whole scene made him uneasy.

"These delays can truly piss you off," he said.

He predicted that the lagging schedule would not only tarnish Orbcomm's reputation, but even worse, it could prevent a major investment partner from surfacing soon. The business depended on launching the constellation before anyone else in the world, but without an $80 million investment in a few months, the lines of customers waiting to purchase terminals, build equipment, or become resellers would dwindle. Annoying questions already being asked about whether Orbcomm was a serious contender left him rankled and anxious, especially when he knew they were still steps ahead of any competitor.

"We've got all this blue sky," he said excitedly. "Think about it—170,000 miles of pipeline in this country with no way to inspect it but by wire or by eye. We've got 13 million shipping containers and boxcars out there, and at any given moment no one really knows where they all are. The Coast Guard has a navigation system with 50,000 ocean buoys that requires regular visual inspection—we could mop that up in a second! Construction companies have hundreds of millions of dollars of aggregate material scattered all over God's creation—and at night everybody drifts off and nobody's left to watch the store. Shrinkage due to theft in the building industry is so common, it's ridiculous. And I'm not even talking about highway projects, grain storage, tank farms, wellhead operations, thousands of independent truckers and fleet operators . . . everyone needs our help.

"I'll tell you something else. The very day we launch the first two satellites, we will have—bar none—the best search-and-rescue system the world's ever seen. And right now, as long as we keep pushing those satellite hackers up the street, we're going to beat every other system into business by six or seven years."

So why was there no significant backer yet? I asked.

Dreher guzzled his soft drink and took a moment to respond. Orbcomm recently had paid J.P. Morgan $50,000 to shop around an investment proposal to a short list of some of the world's most prominent telecommunications companies. But even the analysts with JPM struggled with identifying the market and faltered in summing up the characteristics that set Orbcomm apart.

The basic problem was the Orbcomm "story" still could not be told in a few sentences. Without the benefit of a couple of pithy lines—"Dick Tracy . . . *Star Trek* . . . Flash Gordon"—the system remained a hard sell with investors. With Orbcomm's natural ability to serve hundreds of different niches, the "story" had no single thread, and instead of lending a snappy one-liner quality to a sales pitch, the markets appeared to outsiders as a wishy-washy, cut-and-try concept.

AT&T, in fact, had shrugged them off. France Telecom, Cable & Wireless, and Teleglobe in Canada listened to J.P. Morgan's pitch but remained elusive when it came time to bargain. Telesat Canada invited one of Parker's engineers to give a presentation, but when the young man announced that the total cost for the entire constellation topped out at $320 million, the executives acted insulted, and a few actually walked out of the room.

"These people are preconditioned toward much higher costs," Parker later told his team. "At this price they just don't believe we're real."

Commercial space ventures also scared off many investors because of high-risk, large capital requirements, long payback periods, and uncertain government policies. Most investors wanted a modest return in a short period of time. With undeveloped markets and unproven products and services, it was hard to be convincing. Finally, all the LEO system executives targeted many of the same investors for start-up costs, and when no one knew the Orbcomm name and yet also had Motorola and Lockheed and General Electric knocking on their door, little Orbcomm often looked like an also-ran from the start.

Most recently, in talks with COMSAT, one of world's most prominent satellite service providers for shipping and offshore industries, discussions finally resulted in a series of technical reviews. But no one in the Orbcomm offices felt especially good about having COMSAT as a partner.

"Bad breath is better than no breath at all," someone reminded the staff, but then after one meeting, when Parker asked a representative from COMSAT point blank what the company would bring in terms of cash, the man fumbled and bumbled and avoided the question. COMSAT, Parker said later, "is giving us the bum's rush."

But Dreher shrugged off the travails as petty and natural.

The story needed grooming, he acknowledged, but statistical evidence made the best sales tool. So did the story of another business—cellular phones.

He wiped his mouth with a napkin and reached for the shaker of dried peppers. He had been this way before.

"In 1983 my buddies and I could see that the cellular business looked really promising for rural areas," he said. "People in rural America spend a lot more time in their cars, and just about everybody out there was dying for a better service than what they had. We surveyed something like seven hundred companies in rural America and asked if they used mobile radios, pagers, or two-way radios—20 percent said yes, they use one or the other. You know what the number was, at that time, right here in Washington, D.C.? Five percent.

"Now the same scenario's true for us. There are plenty of statistics—like the 20 percent versus 5—that sell this system. For instance, most people wouldn't believe you if you told them there was this huge market for a data-only service. That's because they're not accustomed to thinking about it in real terms. So you have to make the connection for them: e-mail. E-mail's a data-only system, and it's very popular. You say, 'Orbcomm's e-mail mobile delivery anytime, anyplace,' then people will listen. Everybody knows e-mail's taking over the world. You say data-only messaging, though—who cares?

"Here's another one: Most people think all the money's in urban areas. People know we've already got all these cellular services and mobile phones out there, so they think the whole country's covered. But it's not true at all. Did you know that even though the pager industry claims to cover 80 or 90 percent of this country's population, in terms of land mass, they really only touch a small piece—maybe 20 percent max. Just in the United States, I mean, if you go out to East Jesus, Missouri, there's simply nobody out there to take care of that guy. And it's not because he doesn't want the same service as the guy in Manhattan or that he's not willing to pay for it.

"That's how I built my cellular business. I started out with three buddies. We left Motorola just looking at the market around Denver. In five years we raised $300 million and spread to seven states. By the time I left a year and a half ago, we had a joint venture with a hundred different phone companies; we were providing service to eight million people; we covered seventy-five markets in the rural West and Midwest. I expect to do the same with Orbcomm—but on a much, much larger scale."

Dreher waved down the waitress for a second large Coke. Tall and wiry, he must have been stoking up for a frenzied afternoon. I had watched him knock down the meal with the same alacrity that he gobbled up numbers, with the same intensity he used in honing strategies with his staff.

"The way I sell this system, it's easy," he said. "Most people, you give them the concept, and then they tell you, 'We could use it for XYZ.' They come up with their own applications.

"Next thing you do, give them data. You say, 'Now look how big the universe is of people who would use that application.' We break it down into geographic segments, show them the sales forecast—40 percent returns—and they're sold. It's a little bit of sizzle and a whole lot of steak—exactly where cellular was only a few years ago. Honestly, taking this service to market is not a hard sell; it's a matter of crowd control."

He paused for another bite, another swill, and, I imagine, awaited my reaction.

"Now, let me tell you a little story," he continued. "When I was selling the cellular business, we went to see the New York Society of Security Analysts, a pretty august group that claimed to know a whole lot about telecommunications. I was all of thirty-one years old, and I got up in front of these high-powered, $400,000-a-year security analysts, and I said, 'Well, we're gonna do cellular in rural America.' They laughed in my face. Swear to God—they laughed at me!

"Then I said, 'Well, what do you guys think a farmer does?' They laughed some more, and someone says to me, 'Oh, a farmer drives a tractor, digs a little dirt . . .' 'Bullshit,' I say. 'The man's a commodities trader. He trades commodities, and he needs good, precise information about his markets immediately and often.' Boy, you should've seen that group turn around!

"And that's how you do it. You assemble 'em, give 'em just a germ of an idea, and then watch all their preconceptions fall away. Fact is, you just can't bad-mouth a bunch of huge, untapped markets dying for inexpensive wireless services.

"So we got Westinghouse coming in this week, Teleglobe's knocking on the door, and I'll bet you we'll be doing due diligence with one of those guys in the next few months. They'll see the market pretty quickly after we lay it out there. J.P. Morgan—they were just a little slow on the uptake. I mean, at first I was the same way: The only thing that scared me when I first saw the business plan was wondering whether we could launch all these first few satellites for only $150 million. That's phenomenal. But Parker keeps assuring me it will happen. If we can get a few potential investors over that shock, we'll have a partnership before summer. I expect people will be surprised at who comes to the table."

Dreher tossed a few bills and some change by his plate.

"Let's get out of here," he said, grabbing his coat, already a step toward the door.

Sold himself again, I thought, with a whole lot of sizzle and a whole lot of steak.

"Come on," he said again. "I've got a goose to cook."

And that was how it was done.

The entire Orbcomm project, with no more than a couple dozen engineers, managers, and marketing agents, was not exactly a garage-shop operation, but it was close enough to bring up a commercial business on borrowed cash and still make a run for the heavens before the biggest contenders even finished the first round of design reviews.

At a time when Motorola reportedly sent limousines to cater to engineering consultants at Virginia Tech, the high life for an Orbcomm engineer still meant extra cheese for the pizza. While the mastodons spent untold sums of money on marketing teams to charm wealthy investors around the globe, Orbcomm hired a lonely contractor to make a whirlwind tour of Eastern Europe and sent two of their own through quick circuits across Latin America and the Pacific Rim. Even Mike Dobbles, who happpened to speak Russian, managed to hook a ride for a trip to the Soviet Union and scoped out new markets by interviewing veteran Soviet satellite engineers in historic labs where commercial LEO systems were also being planned.

The LEO frontier was new to everyone, of course, but then everything else was new at Orbital: launching rockets, building satellites, developing markets, managing employees, growing a corporation. Small and plucky, their marketing and engineering teams competed on the most primitive level, like spear-chuckers after big game. They refused to create paper warrens for documents and distribution lists or rely on ICDs and CDRLs—interface control documents and contract data requirement lists—before setting out for the kill. They simply grabbed a spear—a lab bench, a telephone, a handful of 68302s—and lit out down an open path.

Even though competition with the largest players would only increase at every level of their business, particularly as the advent of space commerce became more enticing to traditional businesses, Orbital's young teams were in a race against corpulent legions stuck in the dingy, scuffed-cream fiberboard offices of the past. No one paid much attention when Motorola announced it would pay $700 million for one of the big defense contractors to build 125 satellites for its Iridium constellation. Orbital's engineers scoffed when news spread that Lockheed had entered the small launch vehicle business to compete its own $14 million rocket against Pegasus.

Every business had a different vision, they decided, and Orbital had zeroed in on one it could command quickly. Pledging to build a veritable vending machine business in the sky, Orbcomm's haughty teams believed they were out front and could easily stay at least one step ahead.

Bulletin Board

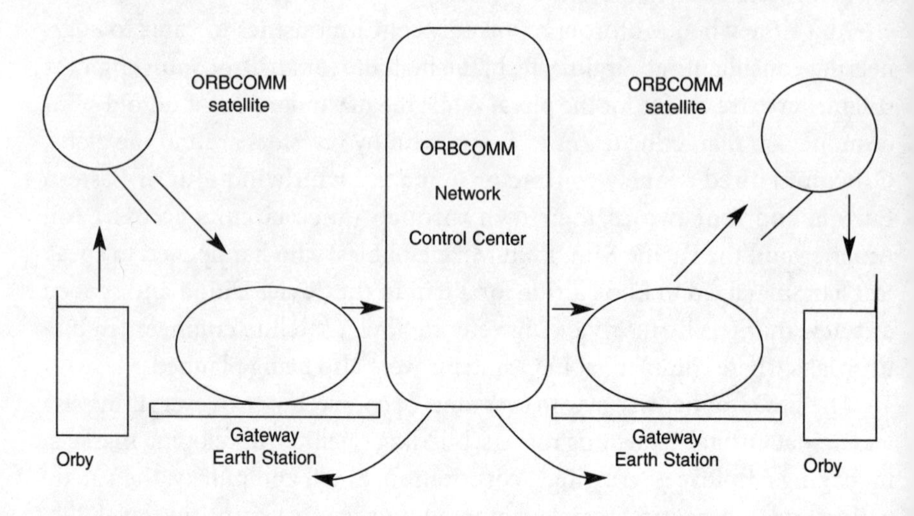

ORBCOMM COMMUNICATIONS

US Commercial Space Revenues (in millions of dollars

	1988	1989	1990	1991
Commercial SATs	550	900	1,050	900
Transponder leasing	600	750	800	850
Earth stations	600	750	850	1,000
Commercial launches (large cap.)	0	150	570	450
Remote sensing	90	115	140	170
Mobile sat services	1,840	2,765	3,555	

SOURCE: US Department of Commerce, Office of Space Commerce and Office of Business Analysis

"What we're doing is basically selling this company for $160 million and we don't even have a single satellite yet! Can you believe it! I don't even understand this business."

—Alan Parker
Orbcomm Inc., president
March 1993

"One shitty review in *Consumer Reports* and we're history."

—Alan Parker
Orbcomm president
March 1993

"The Holy Grail for a space start-up is a quirky niche application."

—Heather Miller
Wired 4.09

"Sometimes I feel like everyone's out to get us here in Boulder. We're a scapegoat for things. We're a focal point. We handle certain functions, like RF communications, but if there's anything wrong anywhere, they point here. For whatever reason, we draw the grief."

—Jan King
Orbital/Boulder
March 1993

"I don't have a bad perception of Boulder. I am concerned about Boulder. I would have closed them a hell of a long time ago, as you know. I simply don't think they're as great as they think they are."

—Alan Parker
Orbcomm president
March 1993

"Orbital Sciences has done great job of building a business and attracting capital, but it has yet to demonstrate that it's a real business."

—Executive of CTA Inc.,
Former small-sat competitor
April 1993

"In this country we are not doing well dealing with our launch culture. Unfortunately, it is based on the NASA perspective of no failure because we've got men on board. And that's carried over to the commercial launch business. The problems are in the culture and we've got to make a drastic change. Look at the number of people it takes to launch the shuttle and you'll see it takes over 1.5 million signatures to fly one shuttle mission."

—Ed Garbris
Space Systems Development, NASA
April 1993

CHAPTER EIGHT

Spring 1993

Elephant Mate

Dave Steffy found out about the Teleglobe deal at breakfast one morning in May, as he flipped through *The Washington Post.* When he arrived at work, his engineers passed him in the hallways with suspicious looks, as if to say, "Why are we always the last to know?"

People weren't so much upset about being left out of the loop as extremely curious about how David Thompson had managed to swing the deal. A number of them had come to work at Orbital to learn the secrets of how to build a start-up in aerospace, expecting to open their own businesses and make their own deals someday. The tactics and subtle maneuvers of the company's executive staff always intrigued them.

His engineers were, generally, still quite happy, blissfully ignorant of whatever went on outside their own project. They had appeared particularly sanguine ever since Bob Lovell and his executive staff moved to the new Orbital headquarters a few miles north of Dulles airport. That event left spacious, empty rooms for them to spread about in and offered further evidence to them that the company was healthy and profitable—no longer tethered to rented office space.

During the spring of 1993, the company had set itself conspicuously apart from suburban Washington's clenched suburban commercial districts by creating a permanent residence, a glassy five-story office building and high bay set on a large isolated tract of rolling countryside in northern Virginia, a plot of land whose name evoked the Orbital image of itself as of a better class, the beau monde, the cream of the crop—*Steeplechase.* With the move to

Steeplechase, David Thompson finally established a mature presence as one of the largest private employers in the Washington area. In the spring of 1993, with the move, the company seemed to announce its ambition to compete in the world of aerospace giants.

The satellite engineers would not take up new quarters on the fourth floor of Steeplechase until later in the summer, so in the meantime Steffy's team claimed deserted executive offices at Sullyfield and filled them with hefty lab benches and tall banks of electronic sensors and monitors. They stationed themselves at odd angles in large abandoned executive suites, stood next to broad windows, and labored in maximum sunlight, like basement artists finally coming aboveground into the liberating freshness of sunny lofts.

In the absence of executive oversight, at one end of the hallway, the team's mechanical engineer, Tony Robinson, pasted a sign saying, "Antenna War Room," and at the other end the operating system guru, Morgan Jones, and the new OSX consultant, Rob Phillip, hung "Do Not Disturb" signs on the doorknob of what was once Lovell's rather luxurious office. On the door itself, they taped another sign: "OSX Cafe."

They enjoyed "casual day" every day now. Without annoying corporate rules, most of them looked like they had just rolled out of bed—and probably had—when Steffy strolled into the coffee room around eight every morning.

After circling the second floor on his usual route that day, Steffy wandered down to the high bay to observe the morning's shock and vibration tests. He watched engineers rap the satellite body with rubber mallets and strap delicate electronic components to vibration machines that simulated the quake of a rocket launch. Then he dropped by the downstairs conference room to inspect the actual "boxes," a new set of golden, glittering, lightweight housings built to hold the satellite's electronics. Finally, he slipped back to his own office to prepare for an important afternoon meeting with Alan Parker, Bob Lovell, and the executive staff.

The Teleglobe deal formed the agenda for his one o'clock meeting. From the morning e-mail, he could see that the contract was not as neatly packaged as CNN and *The Washington Post* reported. Technically, the Canadian company had done nothing more than sign an MOU, a memorandum of understanding, leaving the deal hanging on a series of detailed examinations and reports, including a thorough technical review by Teleglobe's consultants. Persistent questions about who would be majority owner of the Orbcomm business—Teleglobe or Orbital or some as-yet-unformed partnership—also apparently left a sense of doubt and anxiety lingering in Orbital's executive suite. Teleglobe wanted majority ownership.

Considering his attitude about corporate "intramurals," as he began to meet engineers on their regularly scheduled appointments to his office, Steffy chose not to mention any of his own doubts about the partnership.

Aside from being irked at not knowing more about the Teleglobe deal, his engineers still sounded excited about their work, particularly the Freshouts. Several members of the team had put together a proposal for a new satellite called Lightning Sat, a mission that had been under review for several months at corporate headquarters and that Steffy's team jumped on just before Christmas. NASA, they had learned, wanted to pay more than lip service to the new era of "Microspace" by signing more science-oriented small-sat projects that matched Dan Goldin's rhetoric about changing the culture at the agency. Through the industry grapevine Orbital's executives had learned that NASA needed a small satellite to record and monitor the incidence and pattern of lightning strikes around the Earth. J. R. Thompson, a former NASA executive who had come to work for Orbital about a year before, had planted his hooks in a possible contract at NASA offices in Huntsville.

The mission seemed made to order. With expectations that their new satellites would eventually lead a production line for science experiments under the name "Microlab," Steffy's team had built the Orbcomm spacecraft consciously thinking of it as a kind of aerospace Bauhaus design, a multifunctional prototype in a way, that could be adapted to many other kinds of low-orbit experiments in the future. The production of a Lightning Sat for NASA would serve neatly as their proof-of-concept mission for a Microlab bus and could be used to attract interest from the scientific community for small, generally low-cost test platforms, just the kind of test beds scientists had been seeking in space for decades.

There was only one problem. The Orbcomm satellite was much harder to engineer than anticipated. With the launch date lagging into early 1994, the idea of also building a separate Microlab bus for NASA squeezed Steffy from two directions.

He knew his bosses would never refuse the action. A little Lightning Sat would not amount to much more than a $4 million contract, but in recent months, with the unexpected cancellation of lucrative Star Wars rocket contracts during the first quarter and the usual spate of government orders drying up under severe post–cold war budget cuts, the company's Space Systems Division really had no other new business on the horizon. None. Not a single prospect. Although the company appeared healthy, as a whole, Lovell and his staff were panicked by the thought of looking into a dry season—interminably dry, in fact. Steffy never told his team how dismal the future appeared beyond Orbcomm, but he knew the problem was real.

The lightning mapper and another satellite, an experimental weather payload called GPS Met, were actually the only new revenue targets in sight. More than likely Steffy's team would soon start to work on them both—two new development projects—*simultaneously*. Three, in fact—including Orbcomm.

Initially, the extra work sounded simple: Strip out the Orbcomm electronics, knock out one payload deck, mate two Orbcomm bodies together, and—*voilà*—basic transportation for all kinds of new science projects. With GPS Met and Lightning Sat, Orbital could generate revenue but more importantly, sell the concept to the industry at large. If Microlab sales also stimulated Pegasus rocket production, Steffy and the Freshouts could become a revenue leader for the entire Space Systems Division.

"That's the good news," Steffy told his engineers as they met that day. "We're almost certain to win those two contracts—with potential for revenue as far as the eye can see."

"And the bad news?" they asked.

"The bad news is, who the hell's going to run all these projects? I mean, I'm going to have a coronary if I have to build two more satellites by December."

The Freshouts countered that they didn't even have the first two Orbcomm satellites built, they still needed to plan the constellation, and they questioned how he could add two more projects onto their present workload. Steffy's answer came quick.

"Welcome to the leaky boat," he said, a response he would begin to use more and more often as the year went on.

But few actually complained. In fact, when he met with Rob Denton to show off new housings for their electronics—"We cut 50 mil off every shelf," he said. "You won't believe how light they are now"—Denton expressed pure delight.

"Cool," Denton said. "Hey, man, we'll be making these things like McDonald's hamburgers."

At first Steffy looked surprised. Denton always had an optimistic view of things, but the new missions would double his work.

"And in my spare time," Denton added with a smile, "I'll still be building the power system for Orbcomm."

"And in my spare time," Steffy said, "I'll try to get back there to take a look at it."

About midafternoon Steffy reached for a paper bag sitting under his desk, and grabbed a cheese sandwich and an apple he had packed for lunch. John Stolte, his new assistant program manager, drew a chair up to the conference table.

"How are we supposed to react to all this news?" Stolte asked. "We'll be going into double shifts by July just to keep up with the first two Orbcomm satellites."

Before building the two final satellites, the plan was for the team to build two other models—an EDU (engineering development unit) and a Qual (qualification satellite)—with which to conduct the initial tests before completing the flight models, which they called FM-1 and FM-2. That made four satellites in all—six, if they included Lightning Sat and GPS Met.

"The only way we'll make schedule is if we build Qual and Flight at the same time," Stolte said. "I don't think our guys can handle that kind of load. I don't even want to think about the risks we'd take rushing two Flight models into a premature launch on December."

"January '94," Steffy corrected.

"January, whatever," his assistant said.

Steffy bit into his sandwich.

Just at the time that Orbital desperately needed a substantial infusion of fresh capital, it looked as if the stalled deal with Teleglobe could force Steffy's boss to insist on expanding Orbcomm to include two more satellite projects.

"You've heard of the golden rule," Steffy said. "He who has the gold, rules."

"Yeah," Stolte said. "I guess."

"Well, we can't do it any differently until Teleglobe signs the contract for the $80 million. Unless we hire another program manager for these new satellite projects and figure out a way to clone all our engineers to build them, I think all of us will be putting in a little more time. We may be working longer hours for at least a few more months."

"You mean, until Teleglobe puts in the eighty, we're doing three projects?" Stolte asked.

Steffy shrugged but didn't say anything.

His assistant continued: "I'd just like to see the look on Alan Parker's face when he hears that we're going to divide our time on two more satellites—that aren't his!"

Steffy looked at his watch. "I suspect he'll know a lot more in about two hours."

The engineers had already met Teleglobe's executive negotiators and inspectors. With little advance notice the Canadian team had arrived one cool bright morning and nosed around the labs. It was the first Thursday of the month. April Fool's Day, actually.

"You can think of this like the final inspection before closing day on your new house," Steffy had told his team the day before. "Teleglobe goes out and hires a bunch of consultants who know a lot more about inspections and what to look for than the prospective owner does—you know, dead mice in the attic, leaks in the ceiling. At worst they go back and tell Teleglobe, 'The satellite's neat and all, but it's got a few leaks in the schedule.' The deals are actually done at that level, not at your box level. They can always come back and say, 'There's this problem and that problem, and the total system's worth 90 percent of what you're asking.' But I mean, pick three numbers—high, low, and your best guess—and they'll start with one of them. At that point it's just a matter of figuring out how to tell stockholders they're getting a good deal. And it'll happen before you know it—like closing on your new house—only billion-dollar companies still take a couple of months rather than a couple of days."

Everybody felt so confident that Steve Gurney and a few other Freshouts left the lab early that day to scout out a basketball court. A few others stepped out for a beer. Steffy might have joined them but at about two P.M. he was making his way upstairs to the second floor of Orbital's headquarters to join a well-dressed crowd of executives and managers, all of whom awaited Alan Parker's final instruction before the Teleglobe entourage pulled in on Thursday.

It was one unforgettable moment.

"Opening kickoff!" Parker was shouting, gathering the crowd enthusiastically into his conference room. "Arm in the fire, guys!"

While everyone milled around a coffeepot, Parker drilled Steffy about what he expected of the satellite team.

"Teleglobe's going to want to see how deep your bench is," Parker said. "You make sure that every one of your people are dressed in white lab coats. This is no joke, son. Clean up your labs. Put 'em in coats and ties. And when Teleglobe drops by to talk, don't oversell it. Tell them what we're thinking: We may have to press the gates to make it work, but at this point we see no job-stoppers. Got it? You say what you have to say. But don't try to talk past them. These guys are used to big-mouthing from INMARSAT and Telesat. They'll see past it. Believe me, they have hundreds of engineers as good as your best. You won't fool anybody."

He detected Parker's nervousness, but at least, he thought, Parker didn't know the extent to which disastrous failures in the lab during December had pushed the schedule even further out from under them.

Once the meeting began, however, Orbcomm's chief engineer, David Schoen, started to pepper Steffy with detailed engineering questions. The exchange grew heated, and soon the entire meeting was swamped in technical jargon, satellite acronyms, and a slew of data points.

"That's enough!" Parker finally said. "You can dazzle us all day long with your brilliance, okay? But there's really only one thing I want to know: This new schedule shows you've decided to push out our launch to December 12."

For months Steffy had been telling Parker his team would be ready to launch in October. But when he looked at his handouts—it had to be a mistake—he had written down the actual December date, meant for his eyes only.

"It's just a typo," Steffy said. "Here, give me back that copy, and I'll make sure you get the right date."

Schoen snickered: "I think the reason Steffy wants the copy back is so he can burn it."

The fact was, Steffy had adjusted his own private schedule because he had good reason to believe that the Pegasus rocket was experiencing its own delays and wouldn't be ready for launch until December, either. He had figured, if he could withhold the bad news long enough, then the Pegasus rocket manager would have to take the blame for the delay, not him.

It wasn't lying, really—just a little game often played in aerospace, commonly known as "launch chicken."

But Parker saw through the deception, and immediately he started around the table firing questions at other managers—the ones in charge of the rocket, the network control center, the ground stations, and the antennas—until he saw clearly what had happened. Flushed and angry, he turned again on Steffy.

"Give me your real schedule," Parker demanded.

"I'll know more when I go back this afternoon and rework everything," Steffy said.

"He means, when he goes back and 'sanitizes' everything," Schoen sneered.

Steffy shook his head, and Parker, for a moment, seemed as if he would relent.

"All right then. What else should we expect?" Parker said. "Anything else you want to tell us about?"

Steffy thought for a second, drew a breath, and said: "Okay, the schedule's screwed up. Bob Lovell decided to bump launch from October 28 to the twenty-ninth so some of us could take a Sunday off."

The smart-aleck comment passed unremarked. From there on Parker ignored the satellite project manager and went ahead with his overview of the next day's negotiations.

Steffy rested his head in his hands and said nothing more.

The next day the satellite-builders' labs looked immaculate, just as Parker requested. When the Teleglobe entourage came through at midmorning, the hallways stood empty. A hush had descended over the entire Sullyfield office building. By early afternoon Steffy began to receive a regular series of updates on Teleglobe's whereabouts, with news of what was said and where they had gathered.

Work continued in only a few areas. In the high bay a young Freshout knelt down in front of the hanging body of the Orbcomm satellite, which dangled from a crossbar 20 feet above his head. Using a rubber mallet, he gently tapped the body and recorded its resonant qualities on computer. Draped in the blue lab jacket, he looked like a Buddhist monk tapping a gong, calling the devoted to afternoon meditations.

A little after four, executives from Teleglobe and their consultants finished a short walking tour of the Sullyfield high bay, and the Canadians joined Bob Lovell in an empty room just down the hall from Dave Steffy's office to seclude themselves and begin negotiations.

While the Elephants met, Steffy reviewed his team's status reports for the week, seeing time cards with up to 170 hours posted for each engineer during the past two weeks.

They had met the deadline only by turning on the afterburners.

Consequently, the engineers had not seen the first truly beautiful days of the year. In Washington the Cherry Blossom Festival was under way. On other projects engineers were talking about skipping work for the kite-flying contest on the Mall downtown. It seemed like everyone's wife was pregnant—John Stolte's, Gregg Burgess's, Dan Rittman's, Shawn Curtin's—and with the first hint of spring, most people in the company made an occasional break from the office. But Steffy's Freshouts, from the previous weekend until late Wednesday night, lived in the electronics lab, conducting their first "smoke" tests, firing up their EDUs, and watching to see if anything fried as they slowly upped the juice.

Some of them had gone days without a change of clothes, finishing the last two weeks of March with individual work logs averaging 150 to 170 hours. The labs looked like jungles, with black cables running in every direction from computers to subsystems and up to the roof for GPS antennas, snaking across the floor, intertwined with phones lines and wires tagged to spectrum analyzers, climbing across workbenches into the very guts of the green shell body of the first Orbcomm EDU. In the blue glow of their Gateway monitors, fed by the pulse of air canisters next door, where women were constantly building new boards and blowing dust off the surfaces of things, the satellite team kept up a hard-out pace.

On Wednesday morning, the day before Teleglobe's inspectors came to tour, Gurney brought his basketball to work, just in case they finished early. Steffy had given them the okay to leave, as long as they came into work the next morning neatly dressed and cleaned up for the inspection.

On Thursday, as Steffy flipped through the status reports, he thought of their progress over the last few weeks and wondered how the negotations fared down the hall.

Rob Denton leaned into his office. Outside the day had turned dark, the sky rumbled.

A crack of lightning split across the horizon before Denton could speak. They both turned to see gray thunderclouds blinking with light and a sudden hard rain spitting against the window.

"I guess this means the elephants are mating," Denton said.

Only a Freshout would compare an $80 million deal to sex in the afternoon.

By the time the Teleglobe delegation left with the storm, few people remained in the building.

Around six o'clock Tony Robinson wandered up from the high bay with another new antenna model in his hands, looking for someone to join him for a beer. The hallways and labs were empty. For the first time since he'd been at Orbital, the entire team had left before dark.

Quietly, he cracked the door to Steffy's office and, surprised, found his boss standing at the window.

"So do we have a future?" Robinson said.

Steffy looked handsome dressed in his dark, neatly cut suit. Standing against the window, his silhouette looked clean and trim, showing the outline of a fresh haircut.

Steffy smiled. His body seemed to relax.

"You know, Tony," he said, "sometimes it's just really encouraging to work in this place."

The elephants had bonded.

Steffy nearly dropped his baggage when he bumped into a doorway leading into the conference room. In one arm he carried a cardboard box bound with strapping tape, and in the other, long coils of Stanley tape bundled together by string.

"Look, it's Fibber McGee and Molly!" crowed David Schoen.

Steffy grimaced but otherwise ignored the laughter from the executive group. He immediately spread Tony Robinson's first good cut of an antenna model out in the center of the conference table. The people seated around the room included David Thompson, J. R. Thompson, Bruce Ferguson, and Bob Lovell, as well as Alan Parker's group. He untangled the aluminum strips of measuring tape and propped up Tony's device like a potted plant. In moments the room hummed with wisecracks about the model's resemblance to a lava lamp or a bonsai tree, as well as a few more serious and analytical comments.

"You mean you're going to stow that thing as an antenna?" one of Schoen's men asked.

Steffy reached for the device and lifted the strands of tape by pushing on a button like the catch in an umbrella handle.

"'Stowed' is a misnomer," he said. "The good news is, it will probably get us the pattern we need. The bad news is, it would change the shelf layout inside the satellite. We would have to shuffle the boxes around pretty severely to accommodate it."

"How's that thing hinged?" J. R. Thompson's voice boomed.

The gruff tone delivered with a definite southern brogue gave away the speaker's judgment. J.R.'s last job had been deputy director at NASA overseeing the space shuttle's return to flight after the *Challenger* tragedy. He had probably heard more about bad hinges, sticky gimbals, ragged drive assemblies, and worn O-rings in the last two years than he could ever stomach again. That he was prepared to pummel any mechanical design set before him seemed clear. Steffy had come expecting a fight.

"On the other hand . . . ," Steffy said, sidestepping J.R.'s query. He reached for the box and used a penknife to slit open the taped top. "On the other hand, there are no hinges on this."

From the box he lifted Tony's latest miracle: a black graphite quadrifiler, still tied with a string in a nested bundle. "One of our consultants around here told me about six months ago not to get too interested in antennas. Well, I'm afraid I couldn't avoid it."

While he made room in the middle of the table for the quadrifiler, Steffy dropped names of academic papers he'd absorbed and names of their authors—Stegens, Schwartz, Skolnik, Stutzman—which made him sound like the resident antenna expert.

He untied the string and, with only a modest introduction, let the helix drop. It jumped and writhed, uncoiled and flailed itself out across the middle of the twenty-foot table. Executives reached for their styrofoam coffee cups, papers, and briefcases, moving them out of the way of the kinetic coil. No hinges, no mechanisms, nothing apparent in the design to hang up, snag, or disconnect—that message came out loud and clear. Steffy folded his arms and looked around the room.

From the way the executives nodded and hummed, it was plain that the demonstration had impressed most of the group—J.R. notwithstanding.

But as Steffy went into his analysis and then built up a head of steam for a technical overview—information he expected would clinch the sell—Schoen interrupted.

"You might as well save your breath," Schoen said firmly. "We know about your plans. And we're not going to let you avoid it by the little show-and-tell here. I'm happy to see you've found an antenna you can stow, but there's no way you can convince me that this Lightning Sat and GPS Met project won't play havoc with your schedule."

Steffy's face grew red immediately.

Schoen continued speaking, following his opening threat with references to Orbcomm-X and a curt reminder to everyone present that after X died, Orbital's management had agreed they would never rush a satellite to launch again.

"In a small dark room, I'm sure we'd be in full agreement," Steffy said. "But as you see, there are other people here who might have different opinions about how we work this out. The decision is not my responsibility."

"Well, I was told not to say anything this afternoon," Schoen said, even more sternly. "But I promise you, if your engineers are working on anything but Orbcomm satellites, you better believe I'll make sure Teleglobe finds out about it. You're not going to double-talk us out of our agreement to launch in December!"

"Wait a minute!" J. R. Thompson said, jumping in as if the remarks had been directed at him. "Nobody's gonna double-talk it!" The big man shot a mean look toward Schoen and snorted: "Just keep your eye on the goddamn ball!"

Soon the two men were shouting across the table at each other. Schoen insisted that he knew about other high-level discussions between satellite executives, who were apparently planning to claim rights to the first Orbcomm Qual vehicle, the preliminary model used for testing before the final two satellites were built. The plan, which Schoen apparently knew of in some detail, would let Steffy's team and the Space Systems Division strip down the Orbcomm qualification satellite, reconfigure it, and sell it to NASA for the Lightning Sat mission.

"You can't deny it!" Schoen said. "You know the Qual vehicle belongs to us. It's in the contract! And you can't just assume we'll let you sell it to someone else! So don't yell at me. I've been yelled at enough already, and I want to know what you're going to do about the schedule! We haven't even gotten to the hard part yet."

Steffy must have realized then that his friend David Schoen knew everything. In fact, from the way he spoke, every executive at the table knew also. It was as if they had all seen a copy of his latest revised schedule plan, the one he and John Stolte had agreed to that morning in his office. Steffy's best way to survive the meeting would be to keep quiet and let the Elephants storm.

"Come on now," said J.R. "You know all we want is for Dave Steffy to maximize Orbcomm's chances of making this schedule."

Steffy looked strained as the executives continued to talk about his schedule. The schedule set for Orbcomm had always been accelerated—"success-oriented" was the phrase used most commonly around Orbital to refer to schedules. It was as if the company had an unstated policy to bid low and accept the customers' most optimistic schedules, never taking into account vast probabilities that tests would have to be conducted, problems would crop up during testing, and delays caused by redesigns would snare the production cycle. All the program managers at Orbital, not just Steffy, lived under the same executive mandate to meet deadlines that often looked tight at the beginning of a program and later became impossible.

When Schoen started again with J.R., Steffy finally leaped in and fought for himself.

"You know," he said, "no one ever mentions that we lost a month last year because we didn't have enough designers to support all the programs. Or notices that we lost a couple of weeks recently because one of our vendors got hit by lightning, or that the slip in our schedule's mostly due to vendors who have had trouble soldering the boards and putting microcontrollers in at the right angle. It's taken us more than a year just to find a way to build a quadrifiler antenna that will fit the dimensions of a Pegasus—and it's taken nine months to hire an engineer who can do the proper analysis for the attitude-control system. No one ever mentions that this satellite is just a hard problem. And . . . you know what else? It also doesn't help that the brass level in these meetings keeps going up!"

With that final comment Steffy went too far. Schoen shut down. J.R. grunted and laughed. Alan Parker leaned forward and set both hands down in front of him on the table.

Not everyone knew how hard Steffy's engineers worked or the depth of the problems they faced every day. And perhaps they didn't need to know.

Parker took his time.

"So, Steffy," he said, smiling at first, like the best hand in a poker match, "just tell us how much time you've already spent on these other two satellites. What's it been in the past week for LightningSat—30 percent?"

Steffy didn't respond.

"And next week for GPS-Met—40 percent?"

The room was again silent.

"Now," Parker said, "maybe you could also tell us, how much time did your team put on that request for proposals? Couple of weeks? A month?"

Steffy held his silence.

Parker said more firmly, "Are you telling us we can't launch in December?"

Steffy looked around to Lovell, then J.R. They were all three mute.

"Dammit! We're going to tell Teleglobe!" Parker shouted, looking away from Steffy.

"C'mon, Alan!" J.R. said.

Parker turned directly to J.R.

"We can't not tell Teleglobe!" he said. "We have an obligation to let our investors know everything. It's due diligence, buddy—lay it out on the table! And if they walk because you're screwing around with two other projects rather than getting us to launch, you can take your little Lightning Sat and try to float the Orbital Sciences Corporation on that. Whether you know it or not, Orbcomm's the game here now. You got no other revenue. Orbcomm's it, baby! It's bet-your-company time!"

"That's just horseshit, Alan!" J.R. shouted. "You can't blame any of these slips on Lightning Sat. That's just pure horseshit!"

"So what's the schedule look like today?" Parker replied, looking back at Steffy, not for a moment backing down. "The launch was scheduled for December 16. But I understand you're not going to get the electronics from NASA for Lightning Sat until, what, December 10? I assume you're thinking that'll go piggyback with us—so what kind of delay are we looking at? One month? Six months? Come on, how confident can we be about a January launch? We've been telling Teleglobe December's the launch date all along."

This time Steffy did answer. He was almost matter-of-fact.

"To be honest, I'll give us about a 30 percent confidence level for December. The only other option is, we skip building the Qual vehicle, do our tests on the two flight vehicles, and knock off a week or two from launch ops. That's the scenario even without Lightning Sat."

Parker had squeezed his hands until they turned red like a couple of ham hocks. "I can't read the back of your mind," he said. "But this sounds to me like March or April of '94."

"Sorry," Steffy said smartly. "I always wanted to be the only program manager at Orbital not to slip launch."

The way he said it, it broached the sardonic, triflingly close to an insult against every executive in the room.

Although no one had addressed the issue in an open forum like this one, Orbital's remarkable success with Pegasus had created an interesting problem. The company's skyrocketing entry into the aerospace industry had left a measure of accomplishment that set a difficult standard for every subsequent project. The company's accelerated schedules, presented with more of a sense of hope than of reality at times, had earned the company its fair share of contracts, but they had also begun to put a permanent strain on the company's engineers.

"And that thing—that *antenna*," Parker said, motioning toward the quadrifiler model still sitting in the middle of the table. "It looks like some kind of invention to me. Come on, Steffy, give me the real schedule. Put that on the table. Did somebody here give you a launch date just to play with us, or can you tell me what it really looks like?"

Parker sounded angry.

"Our company culture is to aim high," Steffy said. "That's good, most of the time. What can I tell you? I can't calibrate for the change from an aerospace to a telecommunications culture."

The arguments ended there. Steffy set his papers aside, and the conversation turned to other projects and other presentations.

But as the executives began to discuss the company's future in more detail, David Thompson folded his hands together and brought them up to his lips. He leaned forward on his elbows. The exchange had gone on for more than an hour before he spoke. And then Thompson said firmly, in a voice loud enough to stun everyone at the table: "I want to see progress!"

He wanted to win the Lightning Sat and GPS contracts. He wanted all other obligations with all other clients fulfilled—on time. And he expected Steffy to launch in December. He could not have been clearer.

Lovell said nothing. Steffy sat still. Finally, Ferguson broke the silence, quietly and calmly saying what Thompson had, even in his candor, declined to say: "If we fail to meet the schedule or satisfy Teleglobe that we will meet the schedule as it's been presented, Orbcomm has the potential to bankrupt the entire company."

He had said it so simply that it might have been a comment about the weather. But clearly the future of the company was precarious.

At the end of the meeting, Steffy packed up Tony Robinson's fine antennas to take back to Sullyfield, wondering how he alone could keep his team excited about the work that lay ahead.

At midafternoon, as he stepped out into the parking lot, overhead, high above the plush spring greenery of Steeplechase, turkey vultures were slamming into the building, grotesquely casting at their reflections in the glassy facade of the company's executive suite. It was a strange and unsettling sight, one that somehow would always seem tied to that particular day's events and the sudden, dramatic challenge to the company's existence.

CHAPTER NINE

April 1993

The Brain Was an Arse

Morgan Jones had plunged into the second half of his afternoon lecture about OSX when he noticed the new guy take off his wristwatch, set it on the table, and drum the metal arms of his chair like an anthropoid. Everybody around the room people shot puzzled glances.

Mark Krebs, the new guy in the argyle sweater, leaned over a small black notebook and scribbled with colored pens.

To his credit, Morgan ignored the distraction and continued to outline the satellite's message-passing functions on the white board, then circled around the room explaining how messages could be interrupted and stored in buffers for later delivery.

The new guy picked up a red pen and tilted it this way and that. It was as if he had never seen such an instrument before. Then he grabbed a blue pen and drew more hairline detail into a sketch of something that looked like an offensive play diagram for the Washington Redskins.

When Morgan reached the part of his lecture about passing messages between subsystems, Krebs lifted his head. He whacked the table.

"Are you saying I have to yield the floor now!" Krebs demanded. "You're telling me there's no way I can raise my hand and say, 'I want to talk!'"

Krebs pushed his chair away from the table, threw up his arms, and looked Morgan sternly in the eye. Morgan, the team's only expert on the operating system, did not understand. In the land of Orbcomm's satellite software, everyone felt a little shipwrecked, deserted, and washed ashore. Why, Morgan must have wondered, did this guy have to sound off so dramatically? And was he talking about himself or about the software design?

Krebs struck again with a torrent of complaints: "You have created a software task that basically owns the subscriber transmitter, the gateway receiver, the power bus, and my attitude-control system, which means I can only talk when you say I can!"

Then he called the operating system "squishy" and "torturous" and said that OSX left him with a sickening feeling in the pit of his stomach—"formless dread," he called it.

Morgan had seen Mark Krebs's grandstanding before. Since he arrived at the first of the year, Krebs had hollered in meetings, scribbled furiously in his notebook, and in almost every case held in his hand a watch, which he used to announce that any particular meeting had gone on long enough and had reached, in his estimable opinion, the point of inefficient babble. "Enough masturbating!" he would cry, or "End transmission! End transmission!"

A thirty-one-year-old exile from Hughes aerospace, a missiles expert, and not coincidentally a weight lifter, Krebs actually had brought a kind of spirited discipline to the team and more moxie than anyone would have imagined of a techno-weenie from Hughes. He seemed political, like a number of new hires, but just what his game was, no one could say. To stand in his line of fire made any speaker vulnerable to a flurry of prickly queries, jokes, and barbs. Over the last three days of software seminars, for instance, while the lessons took place, Krebs would sit silently for the first few minutes, make his sketches, and then start bombing, forcing the speaker to recoil from an abrupt challenge. He indicated to the Freshouts exactly what a few years of experience could yield that they had not yet acquired for their own professional arsenals: confidence, technical insight, certitude.

Dave Steffy sat next to Krebs at the end of the table and listened quietly while the sudden outburst from this outstanding hire led, as it often did, to quarrels and debates and moans around the table.

Steffy was smiling.

"Hold on a minute," Morgan said. "Wait a minute! Everybody, time out! Let me finish!"

At twenty-four, Morgan looked like a beat poet, dressed in a black turtleneck and leather sandals. He shuffled around the room, talking fast, like Kerouac on acid, though Morgan's drug of choice was caffeine—notably, espresso. Nonetheless, as he scrolled through lines of code, he brought the engineers' attention back to square one, dazzling them again with new details of the evolving operating system.

He had hushed the room, backtracked, and started again.

Morgan returned to the white board, picked up a blue marker, and drew a sequence of bits that looked like a row of Jujubes. As he jotted down figures with one hand, he used the other to gesticulate or smack the marker board, as if trying to make the jujus move, to call them to life. His descriptions of handshakes, acks and nacks, ping-ponging, and "tokening," revealed an ornate but elegant software design and offered the team a half-dozen different ways to code bits that would open and shut doors in the architecture of the satellite's internal communications. His lecture quickly became more impassioned and precise.

The curls in his black hair gleamed with sweaty moisture from his lunchtime run. His temples pulsed. Morgan reached for a muddy cup of cappuccino, poured just prior to the meeting, and took a sip.

"OSX is not perfect," he said. "But it will work."

"This is above my level of caring," Krebs said obstinately. He picked up his notebook and stood up to leave. "I'll be back," he said.

The software seminar was essential. OSX (aka "DOOM") had stymied the team for a year, working less effectively than expected.

The operating system basically served as the satellite's brain functions by snatching messages and dropping them off as it carried information through the tendrils of a kind of nervous system across the entire satellite. The flight computer, where OSX made its proper home, continued to have problems, too. As the simple "brain" of the satellite, where OSX unpacked a regular load of impulses from other internal "organs" of the satellite to be made conscious, the flight computer formed the command center for OSX's primary functions. In a sense the flight computer functioned as the "brain" itself, in which the roots of the nervous system resided and out of which the stringy branches—serial lines—of the operating system spread. Reaching out to many of the satellite's sensors and to all of the "boxes" or subsystems, the flight computer would not so much process all information buzzing through the satellite as it would order OSX to instill strict discipline over communications within the spacecraft and enforce rules that every computer across the system would follow.

The plan to decentralize the satellite's internal communications depended on an elementary design using serial (one-way) lines of communications between the flight computer and a half-dozen other computers housed in boxes throughout the system. It was known as a distributed architecture. According to the plan, the flight computer served primarily to pass messages from box to box and command the operating system. It would set regular times when the boxes (the attitude-control system, the battery control regulator, the transmitters and receivers) could "talk" to each other or snag telemetry data for their own use. It would monitor the health of every box, reset subsystems during emergencies, and, importantly, decide which box could "control" the line of communications at regular microsecond intervals.

Morgan had written his master's thesis about distributed architecture for aerospace systems at the University of Colorado. Although he never intended to carry out the thesis at Orbital, and he regularly reminded Steffy that he had come from school to be a systems engineer, to oversee the satellite system as a whole, and not to be his software monkey, Morgan had already worked exclusively for more than a year on his own intricate plan to connect the flight computer brain to its broader nervous system. Like it or not, he had become the team's OSX guru.

At the moment, however, he was not in the mood to defend OSX from its critics. The decision to use OSX rather than an off-the-shelf commercial system had been made before he joined the team, so he could feign disinterest about it all. But after months of preparation and experiment, Morgan found that he did have a definite stake in maintaining the program. Despite resistance from Gurney and Annette, he insisted that a change to a commercial software would cause more problems than it was worth. Neither of his colleagues was convinced.

So Morgan kept drawing and talking. Naturally, everyone at the table had a chance to comment, and some were especially critical. In any case, a thorough overview of the operating system, which should have taken only an hour or two, dragged on as each engineer demanded more detail and made suggestions for modifying the system. Over the past year OSX had created problems for every one of them.

Sometime at midafternoon, Krebs returned, this time with his new attitude-control system (ACS) team in tow—Grace Chang, Mike Dobbles, and Warren Wu. They took their places around the table and listened quietly.

"The token is only one bit in the control field," Morgan was saying, tapping pieces of a packet he had drawn on the whiteboard. "One . . . two . . . three . . . number four, right here. You can think of this fourth bit as something like the shell in the novel *Lord of the Flies.* Do you remember the conch? Not one of the boys on the island could speak without taking up the conch. Well, the fourth bit is our conch.

"The flight computer hands the conch over to whoever else on the satellite wants to speak, then whoever gets it can use the conch to send a message or ask for data from anyone else. When you're done, you pass the conch back to the flight computer."

Mention of the conch instilled silence around the room. Somehow it evoked perfectly the nature of their own predicament—a group of youngsters deserted on an island with no real leader except the ones who might emerge over time. Grace Chang, who read books by Günther Grass and other people no one else ever heard of, had proposed "the conch" as a metaphor for the OSX token bit not long after she started squaring off against the operating system. It was also the same notion Krebs had attacked earlier that morning during his first outburst.

"Okay, say I'm the flight computer," Morgan continued. "I send out a message to you"—he was looking at Krebs—"*Krebs,* with no conch on it." He drew a line from the packet to a box labeled "ACS."

"The message just goes to you, and you don't talk back. Meanwhile, I send another message to Eric's gateway receiver, but this time I pass the conch." He slashed another line to a box he labeled "Gtwy Rx."

"Now, say I've gone to the receiver with a telemetry request: 'I want to know how many messages you've received in the last 50 milliseconds.' Eric's receiver takes the conch and says to itself, 'Ah, the conch is mine! At last! Now I can talk! I can drive the line!' He goes out looking for the telemetry data, grabs what he needs, then sends it—*with the conch*—back to the flight computer. The flight computer stashes the telemetry into a buffer to download to an Earth station, which will route it back here to Virginia at the satellite control center. Then the flight computer looks to see who's next in line to get the conch."

"So! Just as I thought," Krebs said, immediately thrumming the table with two excited fingers. "Okay, I've got the floor. Let's change the scenario—let's say you give me the conch, and I can talk. I've been waiting to talk forever! I've got a very important messages for the power system—the BCR—whatever—the battery-control regulator. I take the conch because I have an important message to pass to the solar array drive: 'We're drifting in orbit, and we've got to use the sun to slow us down. Clock the solar panels 15 degrees!'

"But all of a sudden something's wrong—the BCR's dead. I've sent a message and the conch to the BCR, but it's kaput. Then what happens? I know that I've sent the conch out, but the flight computer doesn't know anything, right? He doesn't know who's got the conch."

"This is true," Morgan said.

"So if the conch doesn't come back, the flight computer doesn't know to reset the BCR."

"Good point."

"Could be a problem," Krebs said, folding his arms.

Morgan looked puzzled.

"Hey-hey?" Krebs said.

Morgan thought for a second.

Krebs gestured to Warren and Grace and Mike Dobbles.

"Do you care about this?" Krebs asked his team.

The discussion drifted off into the weeds. Morgan's lecture broke out into another debate about how he should modify OSX, and soon the room sounded like group therapy for engineers, with a round of complaints followed by a round of finger-pointing and then more discussion.

Finally, Steffy, for the first time, spoke up.

The project manager pointed out that he had hired a company to modify OSX, and it would officially release the software for the first time in a few weeks. Morgan replied that it would still be months before the company finished the documentation. More than likely, Morgan said, mistakes that appeared in the satellite builders' tests would be used by the company to write a more complete user's guide.

"So you mean we still can't look on page 237 to see how to fix a bug, like you would with anybody else's operating system?" Steffy said. He sounded completely surprised.

Gurney again raised the scenario Krebs had described earlier and found another problem.

"Wait a second," Gurney said. "So what happens when, let's say, you send a command to Krebs? He takes the conch and thinks, 'I need to turn the solar array,' so he sends the conch to the BCR. The BCR's dead, like he said before, whatever. The flight computer never gets back the conch from Krebs, and it says, 'Oh, Krebs's dead, I'll go reset him.' The flight computer ends up resetting the attitude-control system by mistake, and before you know it, the whole satellite's tumbling through space. We're totally screwed."

"I . . . I . . . I . . . ," Morgan stammered.

"Not only that, but what happens if an entire packet gets bumbled?" Steffy asked.

"You're dead, dead, dead!"

Morgan looked around to see who was shouting again. Krebs was smacking his hand against the wall.

"Wait a minute," Steffy said. "This sounds like a shell game!"

"No, there's only one conch," Morgan said.

"Psssss!" Gurney hissed.

"Hot damn! I got the conch!" Krebs cried.

"Whoa! Hold on!" Morgan said.

"Who's got the conch?" someone asked.

"Let me explain," Morgan said.

But it was too late. Krebs played charades with the invisible conch and dramatically exposed the critical flaw in Morgan's design.

He had done it again—exploded into a meeting, collided headlong into a lecture, made observations that led to a chaotic series of questions from everyone present.

Steffy leaned over to Krebs as chatter swelled around the conference table. "Welcome to the leaky boat," he said.

Morgan pushed away from the edge of the table, turned back to the white board, and started to draw a new picture—a "higher order" sketch, he said—of the satellite interior, showing lines for message-passing between all the boxes.

"I'm memorizing this," Krebs said in a threatening voice. No one could tell yet when he was teasing.

"Oh, right, Krebs." Gurney smirked.

Krebs rose up out of his chair and stepped over to the board.

"I'm glad this is your nightmare," he said in an aside to Morgan as he uncapped a red marker.

"It *is* my nightmare," Morgan said.

Krebs started to draw, his red marks crossing—sometimes double-crossing—Morgan's blue, slightly adjusting links between the flight computer and the attitude-control system.

"Now what's the point of the conch if every time we have a conversation, I have to return it to you?" Krebs said. "And what's the point of you giving me the conch to talk to my subordinates when I need to hear from them on my own time, not yours?"

Morgan folded his arms and lowered his head in thought.

If Krebs's observations sounded more personal than professional, Morgan was oblivious. Software logic naturally hid in the shade of abstractions, and the projection of people's personalities onto each satellite box only helped him visualize the "thoughts" and actions taking place simultaneously at the end of every hardwired connection. The critical exchange that Steffy had wanted his team to feel free to hold in meetings like these was, at last, heating up. Morgan's eyebrows were knit in heavy concentration.

Because OSX came in a "roll-your-own" format without any kind of standard documentation, many of its traits remained unfamiliar to the team, particularly to newcomers like Krebs. And because none of the engineers had finished designing software for their own systems, even Morgan had no clue about how to adapt OSX to handle portions of the satellite that still managed to hide in the shadowy netherlands of roughly designed units whose infant personalities remained unformed. Now he had new information.

Morgan finally sat down and let everyone else talk. Ideas emerging from the discussion melded quickly into his imagination, bit by bit. Some abstractions came into relief, and he jotted down notes as he envisioned areas that needed attention. Especially over this longest stretch of time, from December 1992 to May of 1993, as he helped Gurney, Denton, Annette, and Dan pass bits for sometimes up to eighty hours a week, he had needed more clarity about how each box worked. Exhaustively trying to pass one simple message from the flight computer to the receiver and then to channel the same message across the entire system—the "formless dread" Krebs had mentioned earlier—he had created a scheme that was now his very own private nightmare. On nights when he wasn't in the lab himself, he took calls at home from frantic or irritated teammates whose many hours of work appeared lost due to operating system errors. Perhaps he had been so exhausted by the continual labor that he hadn't seen the obvious deficiency.

If Morgan had entered the meeting not quite expecting a melee, once the challenges came, he saw benefits. For months his own intellectual domain was as fractured as a silver moonscape, pocked and gray, an imaginative world that once looked enchanting but lately seemed increasingly cold and treacherous. Comments from Krebs, he could tell, could only help. They would refine his thinking.

Against the accumulation of their long days under artificial lights, Morgan had seen engineers slowly take greater command over their individual boxes, and with their help he found his own understanding of the system growing toward a more mature knowledge. In a way the engineers had become their boxes. Over the last year each person had taken on the traits of a specific subsystem, and strangely they all entered a sequence of separate spheres whose sizes and weights, functions and subfunctions reflected the peculiar personalities of their creators. Whenever people talked about their boxes, they spoke of them in the first person: "*I will send you the conch. . . . I will hammer you if you don't respond to me on time. . . . Give me your health report. . . . These bits I'm sending you are, like, they're my white cells. . . . You send me a heartbeat, okay? . . . Here, I'm giving you the conch. Now you can talk. . . . Tell that 'guy' to shut up. . . . But what if I go insane?*"

Attitude control, battery-control regulator, flight computer, subscriber receiver, gateway communications—each box held someone, as seawater holds a whale, in unusually murky depths. The immersion of their minds and emotions into the confines of a 44-inch satellite felt both unnatural and exhilarating. Sometimes it felt just too cozy. And sometimes, oddly, it felt, when it burst upon him, as Krebs's confrontations had at moments throughout the day, like sunlight, shining with sudden clarity.

Morgan finally lifted his head and went back to the white board to face repeated examinations. Every time Krebs opened his mouth, questions blistered up around the room.

Another half-hour disappeared. Then another. Diet Coke cans, styrofoam coffee cups, and plastic lunch tubs of pasta salad, couscous, and wheat germ cereal lay empty across the table. For the rest of the afternoon, other engineers came and went as word about the conference spread to teammates in the electrical lab and high bay. By six o'clock a dozen remained, some still arguing with Morgan, others with Krebs. The core team—Steffy, Gurney, Grace, Morgan, Annette, and the new ones, including Krebs—had held themselves to the twelve-by-twelve conference room for half the day, arguing over problems they could not resolve. They had not so neatly passed the conch themselves. The discussions finally amounted to a free-for-all against OSX and, in some cases, against Morgan himself.

Finally, Krebs grabbed an eraser and swept the entire whiteboard clean. He drew a new diagram of his attitude-control system, linking it with lines to temperature and light sensors, the solar array drive, the flight computer, and a little box labeled ARSE, an awkward acronym for "attitude reference system." A few people laughed out loud.

Krebs outlined a new scheme for passing the conch, which he apparently had developed privately over the last few days and then refined during his own team's excursions elsewhere in the building during the day.

"I think we're gelling here," Krebs said, finishing the sketch. "We're crystallizing."

Gurney looked curiously at Morgan.

"Morgan, has anything changed?" he asked.

"No," Morgan said.

"Look, I don't know what I'm saying," Krebs said. "This still may be way beyond my caring. I don't think I even understand the concept. Tell me to shut up! . . . But then, let's say I agree with you that on a regular basis I—my subsystem—will be yielding the floor."

Morgan looked at the drawing and slowly came to understand what Krebs had been trying to say to him since lunchtime. For one thing, whenever he spoke in the first-person plural, he actually meant not himself but the ACS, the attitude-control system. Secondly, communications inside Krebs's ACS looked impossibly complex, a design based on timing requirements that Morgan had never seen so intricately presented before. In the box labeled ARSE, Krebs would install a set of equations that would predict the satellite's position every few seconds in orbit, compare it with the actual attitude, and then make moment-by-moment corrections in its own "thinking" (there was no other word for it).

The ARSE had the makings of a brain—strangely enough—a second, separate brain with its own clock ticks, its own message functions, and its own severe timing requirements and feedback loops. To adopt such a plan would force the flight computer and OSX to adapt, as it were, to the presence of another sentient force.

"No," Morgan said firmly, as the plan became clear. "All the interactions must be very deterministic, preplanned. We sit down and say, 'Okay, at 200 milliseconds past the clock tick per second, the flight computer sends . . ."

Engineers attacked again. Morgan later said he felt like the straight man in a comedy routine. Everyone wanted an exemption. He admitted that his plan sounded vague, but the OSX operating system had never been used before. Dozens of his schemes could be worked out only in testing. He handled their comments like the operating system would—winnowing through them, buffering a few, trying to get at Krebs's questions first.

"I don't want to burn you guys out on meetings," said Steffy, finally, looking at his watch.

"Too late," Gurney said dourly, noticing the time. It was almost seven o'clock.

Krebs jumped out of his chair again. But this time he hadn't reached for a marker. He was glancing down at his watch again.

"You're drifting," he whispered.

"But it *is* a systems meeting," Morgan said.

"I know, I know," Krebs said, reaching down to collect his notebook and gather up his pens.

"What? You don't want to see how this turns out?" Gurney said.

"Send me a message in the morning," Krebs replied as he waved goodbye.

The door shut firmly behind him.

Early the next afternoon Krebs returned from Gold's Gym with heat still rising out of his sweater. He went straight to his office, a plain, windowless cube.

I've inherited a nightmare, he thought.

Above his head, tied by a thread to the ceiling, dangled a tiny replica of an Orbcomm satellite, shining blue and black like a quiescent dragonfly.

Occasionally Krebs would reach up and release the model from the ceiling, pinch the long antenna boom, which he had fashioned from a plastic drinking straw, hold it between his thumb and forefinger, and slowly rotate it. He turned it this way and that, flipped it upside down, held it at arm's length while he spun around in his chair, and tried, in his own way, to imagine different orbits.

He faced into it like a father holding his first newborn. Pieces of an equation whirred through his mind. He pictured it in three dimensions, imagined blasting it with the power of magnetic fields and solar radiation. How, he wondered, could such a tiny object like this right itself against the forces of space?

A native of Colorado and a graduate of MIT, Krebs acted like a cowboy, but his thinking advanced in discrete steps, as precisely as a physics professor. He carried the frame of a second-team halfback, hard and bulky in the shoulders, narrow at the waist, with large firm hands and big jaws whose muscles worked fiercely whenever incoming data on his monitor revealed trouble. When he reached overhead for the straw model and twirled it delicately between his fingers, the sight of him caressing the toy was immediately amusing.

Unless one knew the vagaries of what Krebs did for a living—balancing the small satellite in orbit required an ice skater's deft sensitivities to gravity and a rocket scientist's mathematical insight—the sight of such a hefty man pondering a tiny model that looked like Mickey Mouse would leave any passerby in stitches. Serious about his work, Krebs never wondered why people sometimes laughed as they passed his door. Especially at those times when he could be seen alone, bending and swirling like a tai chi student in his office, mimicking the bob and stutter-step of a small satellite reacting to stressful torques, he remained oblivious, so lost in concentration that he never noticed people's reactions to his odd poses.

Developing attitude control on a small satellite like Orbcomm required a specific kind of control from him as well, a discipline Krebs maintained personally and meted out among his teammates with firm authority and clockwork precision. Historically, he knew, most satellite failures stemmed from flaws in attitude-control systems. The system he had inherited when he joined the team in December was rife with them. Unlike every other subsystem on the satellite, his was the one box whose performance could not be replicated on the ground. His tests, which would only occur in software, relied on extraordinarily accurate simulations that he could conduct only by computer.

He should have had other worries, too. Unlike the companies where he had worked before, at Orbital, a single-point failure from a glitch in any particular system could be traced back, in most cases, to one engineer. Although satellite teams elsewhere usually engaged the effort of a large troop of specialists for attitude-control systems, at Orbital the failure of the ACS would leave him a branded man.

On the other hand, Krebs expected that if he found a way to make the system work, the "single-point" issue would leave him identified, almost solely, with one overt and momentous success.

He readily admitted that the challenge to stand out—momentously—was a thought he particularly enjoyed.

Steffy had hired Krebs out of the same company where he himself had first learned the craft of building satellites. Hughes Aircraft Company, through its internship programs and historic connections with the best engineering schools in the country, made an ideal source to satisfy Steffy's increasing need for experienced hands.

Krebs was not the only hire he had made out of Hughes in the last year. Young, talented MIT aero/astro graduates like Krebs (class of '83) often wound up at Hughes, following career patterns of older alumni who had pioneered satellite design and production since early 1960s. Groundbreaking work (the first geosynchronous satellite came from Hughes) had made the company a legendary world leader in Space Age communications.

In the small universe of aerospace engineering, Hughes provided world-class training at one of the oldest and most celebrated houses. With its global influence and remarkably successful Early Bird, Comstar, Telstar, Anik, and Intelsat satellites, Hughes could boast that it was the world's most prominent company in the satellite communications business. Hughes was an empire, a kingdom, a wealthy and historically important corporation. Outsiders admired it. Orbital's founders benchmarked its technical excellence. Krebs loathed it.

He had come to Virginia slouching out of Hughes nearly bankrupt and disillusioned but not yet corrupted. Nine years in California's once-celebrated, now-slumping aerospace corporations had left him cynical about the industry as a whole. As much as he enjoyed building missiles and playing weekend sportsman in California's golden climate, the thought that he might ever become as embittered or demoralized as many of the corporation's older engineers alarmed him. Layoffs and buyouts, which buffeted the industry like passing hurricanes, suggested a cataclysmic change in aerospace by the early 1990s. From the West Coast, Orbital looked like the promised land.

In December 1991, for example, exactly a year before he came to Orbital, the state of California estimated a loss of 60,000 aerospace jobs for the year. Economic studies showed that half the aerospace workers in California laid off in 1989 were still unemployed, and most of those who did find jobs held low-paying service jobs outside the field. By the end of 1992, 19,000 more layoffs pummeled engineers at Hughes and McDonnell Douglas. Of those laid off in 1991, only a handful—16 percent—held jobs of any kind. Rockwell, TRW, and Northrop, where Krebs had worked on missile systems for several years before going to Hughes, had just dumped thousands of employees already—a "correction" during the Reagan defense-budget boom years—and announced that further "restructuring" in the government's defense sector would threaten livelihoods for years. The last days of the cold war turned into an especially cruel season for companies that depended on hefty government revenues.

The prospect of continuing decimation heightened Krebs's concern. Paired with the internal politics and convoluted scheming he witnessed among some middle managers and executives at Hughes, the absurd realities of traditional aerospace became too much to bear. Even as men and women lost their jobs daily, Krebs saw middle managers and foolish executives still attempting to create and expand personal fiefdoms using money from shrinking coffers of government contracts. The business looked obese and insatiable from the inside, even as financial drought set in. Krebs saw his future in the business bounded by the inanities of corporate politics and byzantine relationships.

"The setup there was feudal," Krebs told me one day. "Traditional lines of reporting were carefully drawn and religiously adhered to. It meant that someone could do local optimization of their own career at the expense of the company, no matter how much something cost to do a job. You knew it would always look good if you could get ten times too much money and hire ten times too many people and accumulate ten times too many hours for any one project.

"The group I was in—missiles—was fat and profitable, and my project manager, who I was doing controls for on a navy missile program called LEAP, was great, but he was a technical guy. He was not into the creed of a maximized career at the expense of great engineering. I think most of us were the same way—technically good but politically naive. And at Hughes that left us vulnerable.

"So instead of seeing our group growing on the success of our missiles, we made so much money that after General Motors bought Hughes and made it a subsidiary, the corporation perceived us as a cash cow and drained the money out. At the time there was so much money still coming in for Maverick missiles—that was the main missile used to kill troops in Desert Storm—that for a while people in my division just didn't see what was happening. A huge cash flow from past contracts flooded the place. Twisted management structures across divisions sort of recreated themselves and somehow flourished by divvying up what was left of our profits without any requirement for people to produce or succeed at anything. Suddenly, when GM started to extract more of the booty and Star Wars petered out, there was no new business.

"For a while, my job was to come up with new ideas, make ten view graphs and sell the ideas to people outside the company. But then my boss quit. He got fed up. And since most everything else about Hughes sucked—it was a mean management that didn't care about its employees and, for the most part, didn't even care about the company—I couldn't wait to get out."

So when Steffy called him at his $200,000 home in L.A., Krebs immediately took an interest. Married but childless, he looked back east and saw Orbital's offer as a way for him to extend his engineering skills for many more years. To build something great with his own hands sounded much better than an inevitable move into ponderous middle age and the muck of program management elsewhere. If he stayed in California, he feared he would become like everyone else—"focused on your wife and children, grinding for some giant aerospace company."

In almost every instance, when Steffy called his sources inside traditional companies to snoop for experienced engineers, he heard the same echo of complaints. At Orbital, Steffy would say, an engineer could still influence the growth of a company. Although Orbital's potential still far exceeded its profits, no one there experienced the fear of settling down too early, becoming bored by the work, growing old before their time, or giving in to a feudal system.

Krebs put his home on the market and moved immediately to northern Virginia. He and his wife set up house in a tiny no-frills apartment near the office. After a few months the house in California still had not sold, and the expenses drove Krebs to the brink of financial collapse. Given the sluggish real estate market in southern California, he considered declaring the house a loss and filing for bankruptcy. Some days when he came to work now, the struggles of the Orbcomm project felt like a fight for his life. Only for Krebs, such a predicament seemed like a thrilling prospect.

When Grace dropped by at two, Krebs was still flicking a finger at the little satellite model, wondering why Mike Dobbles wanted to increase the number of thrusters on the satellite from one to two, since pulsing them simultaneously through the satellite's center of gravity would cause the spacecraft to tumble. Dobbles had been bugging everyone about attitude-control problems for months, and now he wanted to destabilize it further?

Grace saw pages of colored drawings from Krebs's notes strewn across his desk and instructions that he had penciled in demonstrating how the ACS would "talk" among its various components—an Earth horizon sensor, Dobbles's propulsion system, two magnetic torque rods, a magnetometer, and a box he had highlighted as ARSE and "Kalman filter."

Taking control of the nightmare he had inherited, Krebs had gathered probably the most organized system of notes on software and electrical design of anyone on the team. He had collected academic papers about passive attitude-control systems and queried the world's leading experts, Peter Josephs and Jim Wertz, about alternatives to the Orbcomm design. And yet on some pages Grace saw odd instructions—"*Bullshit!*" and "*Give me that conch!*"—which he had posted like thought-bubbles in a cartoon strip.

"What did you think of yesterday's software conference?" Grace asked.

"Morgan correctly perceives that things are not proceeding nicely," Krebs said, setting his model aside. "Why? What about you?"

"I was appalled," Grace said.

"I think everyone was appalled," Krebs said. "Everything's all fucked up. Why shouldn't we be appalled?"

"Yeah, but it's taking too damn long to admit it," Grace said.

Krebs stretched his neck over the top brace of his chair. "We used to have the same kind of thing at Hughes," he said. "But there it was even worse. All the REAs—"

"Acronym control," she said.

"Okay, responsible engineering authorities—they would call a meeting with about fifty of their minions to find out what the trouble was. The meeting would last all day, and all anyone would do was to stand up and tell some quaint little story about their technical problems. The good thing here is, at least when someone stands up to talk, everyone listens. At Hughes, while one person talks, everyone sleeps."

Krebs got out of his chair and stretched—madras shirt, brown jeans, eyes bloodshot from overwork. Grace climbed atop his desk and drew her legs into a sitting position.

"I also want to know what you are going to do about our contract problems," she said boldly.

"The Earth sensor?" he asked.

"Yes, the Earth sensor," she mocked. "Their guy, Ass-hammer"—Grace sometimes twisted the names of people she didn't like, such as this man, whose name was something like Eisenheimer—"Ass-hammer changed our performance specs without telling us. Now they're acting like we agreed on a spec."

"Must be some area of physics they haven't figured out yet," Krebs mumbled.

"But we'll fight them, right?" Grace demanded.

"No, we'll cave," he said. "That's my main contract strategy. I always cave."

"I knew it! I hate that! You're such a wimp, Krebs!"

"Hey, I don't have time to fight. We have to find out what kind of algorithm we need here. I have to write a buttload of code."

"So let me see how the system looks today," she asked, squinting as the simulation tracked the route of an orbiting satellite across his monitor. "Real time is so damn slow," she said.

Blue, yellow, and red lines, showing the current dynamics of the satellite in yaw, pitch, and roll, marked the course and inched across the screen.

"You could say the vehicle flies pretty good if it knows where it is," he said.

Krebs looked at Grace, who was obviously puzzling over his response.

"Don't worry," he said. "Be happy."

She balled her hand into a fist like she might hit him, but Krebs had something else in mind. He handed her a page from his notes.

"Here," he said. "We need to know more about Dobbles's thrusters. If he's going to use two thrusters, he'll kill us unless he can fire them right through our center of gravity. If his little mouse farts muck with our ability to point the antenna and solar arrays, we're screwed."

Grace took the paper, scanned it. Krebs tapped his watch.

Any meeting with Krebs rarely lasted more than five or ten minutes. He always had his eye on the watch. With hardly another word, she left to find Dobbles.

Krebs turned back to the simulation. It was done. He tapped his keyboard.

"Survey says!" he shouted.

A set of numbers appeared in a small black window on the screen. He blinked and drooped his lower lip.

During every moment of the orbit, the ACS had exerted good control over two of its three axes. But control over the third axis, which would keep the antenna pointed directly at the Earth, eluded him entirely. He stored the data and pondered his predicament—*like trying to move an elephant on ice skates by blowing at it through a straw.*

His jaw tensed and pulsed.

Krebs reached for the little Orbcomm model, kicked away from his desk, and rolled out into the empty gray center of his office, as Spartan as a jail cell. He stared at the little model and moved his feet to slowly spin himself around in the chair.

If he couldn't make the satellite any larger, he would have to invent something to make it stable. Something inexpensive. Weightless. His ACS would have to capture the power of the Earth's magnetic fields and use them to his advantage.

It would work like karate. It needed to settle paradoxical states with Zen-like authority. It required a higher intelligence.

Krebs looked up at the ceiling as he spun.

A magical moment!

He needed it badly. Somehow, in the process of calculating and gathering the magnetic forces of the Earth while in orbit, his system would have to predict, moment by moment, those forces that could destabilize the satellite and then make some corrections in its own internal magnets to sway those same natural forces to the satellite's own advantage.

How to turn push to pull, how to make a punch feel like a boost, how to fool the Earth's magnetic field into giving the satellite control over its third axis just at the moment the field began to assert control against the indefensible satellite? How could he do that?

He would define it mathematically. Only then could he conduct war against the laws of physics and create a magical moment out of a mass of algebraic problems, statistics, and probabilities that would yield a lower order of artificial intelligence that would fly.

Krebs stood up, shut his door, and returned to the computer.

Moving the blinking cursor around a long set of equations, he tapped in a series of numbers and stared at the screen. He hit delete and started over.

At some point the magic would appear, and he would capture it. Just at the moment when he had waited long enough for the parameters in his brain to click off and open a channel to something he hadn't yet accessed, a serendipitous solution would come to him, and he would quickly distill it to a set of equations. Using his own quirky intelligence, he would apply the final algorithms to his so-called satellite attitude reference system, the ARSE.

In Krebs's crazy world, you see, the brain was an arse and its arse was its brain.

But of course.

His quest for the perfect algorithm had begun.

CHAPTER TEN

May 1993

Wonderboy

"Okay then, you take David Thompson's job for a while and tell me what would you do!"

Steffy's boss threw up his hands in exasperation while the rest of the project managers, in their pinstripes and suspenders, turned to see how their colleague, having pierced the boil, would respond to a challenge that might just have been addressed to them all.

Would Steffy layer his complaints with a soothing balm, or proceed as he had begun, with more invasive surgery?

John Mehoves was waiting.

"No, John, you don't get it," Steffy said, surprisingly direct. "This is not a problem about decreased benefits. It's not about no more free sodas. What you don't see is the whole picture. Bring it all together, people are mad about the lost benefits, they're mad about the lousy pay raises, they're mad about the hiring freeze, the long hours, the unreasonable schedules."

Mehoves, Bob Lovell's division vice president, listened attentively. He had called every one of his managers into the conference room to plead with them to return to their programs with a heightened spirit of mission. The outlook for new work appeared dismal, and if existing projects failed to meet deadlines, the division might sink.

After hearing the vice president's opening speech, all the project managers had sat in stony silence while Steffy went ballistic. Despite Mehoves's challenges, Steffy would not relent.

He kept pounding: " . . . And it's also a case of being an engineer here and you have a buddy in a parallel job at Hewlett-Packard who got a ten percent raise this year when you only got three. There's a lot of that."

"I don't know how to help you with that," Mehoves countered. "If we had given a ten percent raise last year, the company would have—"

"But that's a fundamental misunderstanding, John," Steffy said. "David Thompson would be insulted if we compared this company to Martin Marietta or Boeing except at the time of the year when they do the salary survey. But that's the exact time of the year when he shouldn't compare us, because the competition for our people is not aerospace. Really. Not anymore. Our competition is Sun Microsystems and Intel and Microsoft, okay? In one case I'm dealing with, the competition for the engineer is literally Hewlett-Packard. Maybe you could say that's fundamentally a different business—and I don't think money's our people's primary reason for getting out of bed in the morning, but at some point—"

"C'mon, Dave," Mehoves said, smacking his fist into his hand. "You gotta be realistic. This is all about numbers. You got people. You got rent. You got depreciation on equipment. You got contracts, materials, salaries—what else really matters? What do you want? Tell me what you want. Fewer people? Decreased benefits? What?"

"What we *don't* want," Steffy said, "is to get ourselves in a situation competing for people who have better benefits and better equipment and a nicer salary package. I'm telling you that's who we're competing for."

Mehoves turned his back and lowered his head. No one else in the room had dared intercede. But with Steffy taking charge, everyone could see the time for candor had come, at last.

Mehoves had been with the company since its earliest years. Straight out of the Sloan School at MIT, his role in the creation of Orbital had left him with a decade of memories of lengthy road trips, skittish investors, the company's first NASA contracts, the initial partnerships. He lived with the impression that the same ethos of sacrifice and bare-knuckled struggle still revved the troops. He believed his own lore.

Mehoves looked up again and said: "Well then, if we in the space business can't compete for the best employees, what's the answer? C'mon, Dave, tell me what is it you'd do. We've got a numbers problem. The latest survey shows morale here is slipping across the board. The only revenue we can hope for anytime soon is this Microlab contract. What's your idea? You want to have fewer people and pay them less? You want to offer them less equipment to work with?"

"No," Steffy said clearly, immediately. "Here's what you do: Stop being so aggressive with these contracts."

A few others began to talk. There wasn't a person in the room whose project could meet its intended launch date during the next year. As a group, the project managers also believed the executive staff consistently backed them into a corner by pushing for preposterously short schedules—"success-oriented" schedules.

At the same time, Mehoves was right—the division desperately needed revenues from the Lightning Sat mission and potential follow-ons if the spacecraft proved reliable. But to win the contract, at least one of the managers at the conference table would have to sacrifice at least the next year of his life racing toward an impossible—"improbable," Mehoves corrected—deadline.

Finally, one of Steffy's colleagues, Bob Lindberg, spoke up: "I understand it from your perspective, but the real point is, how do you make this taste good to the guy who's got to do the job now?"

"Okay," Mehoves said, angling down to look Lindberg in the eye. "But first you tell me: How do I make it taste good when I give him a pink slip?"

Mehoves stood up and waved his arms imploringly to the whole table: "Hey, guys, the only thing we've got on the horizon at all are these Microlabs. J.R.'s got a line going in for us to launch two Microlabs a year for as far as we can see. We've gotta stick our nose in it now. And unfortunately we've gotta make somebody here the sacrificial lamb so we can do that, okay? We either do it that way, or we wait and go headfirst against every other competitor in the business. Right now we can win it with only one flight. Tell me, would you bid it or not? And don't tell me you don't want to do it because it only makes our mountain a little higher."

They had all been running up that slope for six or seven years. No one in the room could keep clocking sixty hours every week. Production schedules for all the development projects—Seastar, Apex, Orbcomm—had reached a bottleneck.

"Who has the staff to offload onto another project?" Lindberg asked.

"I know!" Mehoves shouted. "I understand the problem! But you know the game around here, and you know we've got to go through with this."

"I don't think so," Lindberg said, raising his voice, too. "And now you tell us why you would let J. R. Thompson go sell something to Marshall Space that we can't deliver. I would have told him to structure something that was feasible instead of something that just reflects David Thompson's model of what's possible."

Again the swell of multiple conversations rose in the room. Speaking above them, Lindberg hung on to their boss like a bad dog.

Lindberg said, leaning forward across he table: "So tell us, John, which one of your program managers told you we could build another satellite in eight months?"

"Dave Steffy said it," Mehoves said.

"Sure, and you know what? He didn't dare say anything else," Lindberg said.

The other project managers laughed.

"Look, I didn't tell him I would cut his head off," Mehoves huffed.

"No, I guess not," Lindberg replied. "You simply told him you'd feed one of us to the lions instead."

"I only told him we had to win this thing," Mehoves said calmly. He sounded now as if he might be backpedaling.

"I know that," Lindberg said, "but you have to be very careful about what you sell. When you sell it not caring what you do to your people or to your profits in other areas like—"

"I guarantee you the company will end up with a net positive input," Mehoves insisted.

Lindberg again: "Sure, and it'll accrue to somebody's bonus, I'll tell you that! And it won't be ours, will it?"

Steffy might have had a natural interest in the rest of the conversation since it had been his job to initiate the Microlab project with his engineers. But he excused himself, saying he had a conference call with an antenna manufacturer. He quietly slipped out of the room, grumbling, as an uproar began.

David Thompson and his wife, Catherine, had a bottle of Chardonnay chilled by the time I arrived. Surrounded by woods, their narrow two-story home, apparently custom built to blend into the trees, looked more like a townhouse than an executive's manse. Although I also lived in Reston at the time, I had never seen their neighborhood, a secluded spot in the northern part of town. The street to their house was marked with signs saying, "Private Road."

Their cat, Orbit, nuzzled my pants leg as soon as I walked through the front door. The one large downstairs room melded kitchen and dining area with the den, and all three blended together beneath the crown of a cathedral ceiling, an architecture intended to create the illusion of spaciousness. Their home was really quite small and, I noted immediately, modestly furnished.

"We don't entertain a lot," David explained, motioning me to the den while Catherine completed dinner preparations in the kitchen.

"I'm just the average Joe," Catherine explained apologetically. "Most of David's friends are aerospace all the way, but frankly engineering and science aren't topics that particularly interest me. I don't even like to watch the Discovery Channel—that's the only thing David will watch on television."

I noticed the walls of the den lined with oversize books, histories, biographies, nonfiction tomes about engineering, science, and aerospace. David smiled and quietly uncorked a bottle of white wine while Catherine made small talk.

As she moved around the kitchen stove, she let me know immediately that she was not inclined to spend the evening discussing the vagaries of orbital drift rates or the influx of rocket engines from the Russian stockpile onto the world market. David's friends did have that compulsion, but it wasn't hers. She thought of her husband's fascinations and talents as a curiosity—"a gift," she said—one she admired but never fancied. Travel appealed to her much more than rocket launches. Until January of this year, she had worked for a cruise line arranging world tours.

David listened without interrupting, occasionally nodding or laughing as she highlighted their differences, an element of their relationship they apparently enjoyed for its balancing effect. (At MIT, I had heard, he and his steady girlfriend had shared all the same interests—they had even planned to train together to become astronauts.)

When we sat down in the living room, I noticed an orange light emanating from the hearth.

"Oh, the neon sculpture," he said.

"'Wonder Boy,'" she said.

The light, shaped in a sharp zigzag like a lightning bolt, cast a luminous glow against the fireplace and reflected an intense blur of color into the room like a lightning streak crackling inside a cloud. "Wonder Boy" referred to a baseball bat used in the film *The Natural*, the story of a washed-up ballplayer who manages a comeback and rekindles his career by smacking a prodigious home run with a bat honed from a lightning-struck tree.

The piece came from an earlier time in the Thompsons' household. In the few days prior to the first Pegasus rocket launch, as observers ballyhooed Orbital and its preposterously ingenuous rocket—*The Wall Street Journal* published a damning story at the time, suggesting that the company itself was a farce—David had fended off skeptics with a line from *The Natural*.

"We'll let our bats do the talking," he would say.

After the startling success of that first launch—"a home run," they called it—when the flash of that rocket captured world attention, Catherine recalled the remark and sought out an artist to fashion a neon lightning bolt as a gift.

Catherine didn't seem to mind telling the story once again, and for a moment, the room was lit by the memory of a happier time. The neon burned like firelight, glowed like warm coals.

For more than ten years, the company's life and their own had shared a common history, bound them in an alliance against the risks of youth, perils that seemed no less startling now with the passage of time but did reveal much about the importance of luck and bluffs and, at times, the feckless—or fearless—gambles they took.

During a time when most of their friends and many of the company's employees started families, David and Catherine watched the corporation grow with exponential speed, always within their care, commanding their attentions like an infant. Unfortunately, those duties prevented them from enlarging their family beyond the corporate one.

"It's just the two of us here," David had said, almost apologetically, when he met me at the door. "We have a cat, but no children—yet."

Although neither said so, I imagined the corporation's transition into adolescence had left them an opportunity to grow, finally, during the last stretch of Catherine's childbearing years. The fact that she had quit her job and David's comment about not having children "yet" suggested to me that they either already harbored news that she was pregnant or plans were under pursuit.

As we sat down to supper, a small feast prepared with a gourmand's flair, they both talked wistfully of the early years together, when any accomplishment was cause for celebration and the "family" included only a few "young space nuts," *The Wall Street Journal*'s infamous label for the start-up.

A little more than ten years ago, David and Catherine had met in Huntsville, Alabama, where Catherine's father was stationed in the military and David held a job with NASA. They had fallen in love, and when he accepted a new job with Hughes in L.A., she joined him. They latched her red Ford Pinto onto the back of a U-Haul truck and left to live in an apartment with David's friend from Harvard, Scott Webster.

Catherine remembered having to hire the woman next door to answer their phone, a couple of dollars for every call.

"That still sounds like we were paying her too much," David said. "But I guess we weren't getting too many calls for our ideas back then, were we?"

They laughed a lot during dinner. Their dreams had come out of a sense of youthful exuberance and a reckless disregard for tradition—"serious insanity," as David said—but without some of the advantages of youth. In fact, Catherine said, being young had never benefited them. The fact that David looked younger than his years, and more handsome than anyone so smart had a right to be, played to his disadvantage in an industry dominated by grizzled graybeards and gruff, overcompensated former air force commanders. Rebellion was one thing, quite tolerable in aerospace, but to look like a young pup created an image entirely without advantage.

"They're especially hard on David because of his age—still are," she said. "He'll turn forty next year, and it couldn't come too soon. Being young has not been easy."

It might have been much easier, for instance, if he had entered the field with the age and demeanor, say, of J. R. Thompson, a NASA veteran with little apparent regard for social graces and enough years behind him to straight-out strike the fear of God into people. J.R.'s booming declarations—"Well, goddamn you, this is the way it is, this is how it is, and this is what you're gonna do 'bout it. Don't try pushin' me around, you little sumnabitch!"—could stop a charging bull. Not that David couldn't hammer hard when necessary, but the image his youth projected—untried, inexperienced, unproved—put him at a definite disadvantage among the industry's executive set.

Catherine was quick to defend David, too, from periodic attacks from the press, which seemed to celebrate any downturn in Orbital's business or rocket failure, casting fate with front-page headlines and story placements usually reserved for more significant news.

"It's the survival of the fittest," she said. "They've been waiting for him to fall. It's the way they test you. Sometimes it's so bad, I think I'll send out announcements when he turns forty and say to the world, 'Okay, David Thompson is forty years old. Now get off his back!'"

I recalled for them the conversation that afternoon between John Mehoves and the project managers. David said he had taken a call shortly before I arrived and Mehoves relayed the story. He sighed wearily.

"I'm very concerned about the morale problems," he said. "The business is just not as much fun as it used to be—not for anyone."

In the early years each person in the company understood the risks and challenges of a start-up. In those days it was common that news of some technical or, more likely, financial problem left them staring down the barrel of disaster. Their work often occurred under bet-your-company pressures. They had never nailed enough contracts at any one time to assure five or ten years of work, like most aerospace companies, so each year was a gamble. Anyone who had been with them since the beginning or at least since the Pegasus project would have understood why Orbital set its bids at numbingly low rates. The underlying ethos always implied that the company would sustain losses to make an initial sale and then make up the difference by asking people to work a bit harder, to put in longer hours and design more efficiencies into their technologies.

When a team of thirty or forty engineers comprised the entire payroll, the pleasure of sacrifice had been its own compensation. With the number of Orbital employees now growing to more than 1,200 and revenues close to $175 million, it seemed as if those expectations had changed. At the same time, David and his executive staff still faced down the barrel of disaster, they worried about cash flow from quarter to quarter, and at the moment they were searching for partnerships, mergers, and potential acquisitions to raise them out of the fire. But he couldn't say that to anyone. He couldn't expose their plans to the competition or even assure the rank-and-file that everything would improve or even change soon. The fact was, he didn't know what would happen next.

"Raising money is hard" stood out as the one phrase that rang so consistently through the first ten years of Obital's life that it could have been the company motto. Ironically, as Orbital's success allowed programs to grow and succeed, every new venture still resembled the first in terms of risk and the need for capital investments.

David said he understood that his more veteran engineers and managers had grown tired of staring down that same nasty barrel. At the same time he knew that Freshouts and other new hires had grown accustomed to the corporation's comfy, well-equipped offices, and, recalling the company's hot reputation, which was probably what attracted them in the first place, they probably wondered now why they still had to work nights and weekends, why Thompson had clipped their insurance coverage, and why the company's bright image didn't match its tepid stock price. If Orbital's executives still expected that someday the company would be the best in the business—and Thompson had no doubt that it would—new employees had come to expect that to translate palpably to their personal advantage. They wanted stock options, beefier salaries, and more benefits of all kinds for them and their families.

Aberrant expectations, David said. The entrepreneurial spirit that once drove the company, that still existed in the minds of some veterans like Antonio Elias and Dave Steffy, was lost on most of the new people—and maybe even on some of the older men like Bob Lovell and Jan King, too, and on most of the project managers, as everyone grew increasingly weary of never-ending financial struggles.

Catherine finally interrupted David as he talked.

"I keep thinking that if you could just bring them closer together, they could see that you're sincere," she said. "Once you're all in that big new building, they'll see you more often. You'll be more like a family again. People will see how much you care."

"It's just damn hard to satisfy all the demands," he said.

She heard him, then looked at me, as if she knew what I was thinking—what many of Orbital's employees suspected.

"He's not in it for the money," she said tartly. "It's never been about making money. They have to prove something to investors at the end of every quarter. But he's not in it to be rich. I mean, here's a guy—you can't even buy clothes for him on his birthday!"

He worked nearly every Saturday, she said, and the door was always open to anyone who wanted to come in to talk. That had always been his policy—actually, it wasn't even a policy, nothing so formal. So why didn't anyone drop by?

He couldn't explain it. Occasionally some of the old gang would—Elias, Ferguson—but rarely anyone else. The cynicism he detected creeping into the ranks—people calling the new office building "the Taj Mahal," for instance, and rumors about him purchasing a $40,000 marble table for the board room—made a troubling statement about the widening chasm between the company he had wanted to create and the one Orbital was becoming.

The company had to grow to survive. And it would keep right on growing—he hoped.

His wife's observation that the employees didn't know him, that he needed to bring them together, sounded quaint and innocent, a reflection of that time when Orbital really was a mom 'n' pop space company.

"It was just a couple of years ago," he said, "when we could order out pizza and everybody could meet right outside my office for lunch."

But that seemed foolish to think about now, indescribably naive, a memory just as well forgotten.

Most likely, the changes at Orbital were nothing more than a natural outcome of the company's struggle toward maturity. At some point, a corporation reaches a stage in its development when rumors can't simply be met or laid to rest with a direct conversation with the chief executive over a pizza lunch.

"Unfortunately, they don't teach you much about 'morale' problems in business school," he said. "It's not teachable. Feelable, maybe, but not teachable."

Our conversation reminded him of the night before the first launch of Pegasus, when he and Scott Webster sat in a motel room in Palmdale,

California, worrying about how the recent story in *The Wall Street Journal* would affect their IPO, which hinged on the success of the launch.

He remembered sitting with Scott right across from him on the next bed, both of them staring into space, both mired in the same thoughts: *What have we done? We've come this far, and now* The Wall Street Journal *is trying to kill us.*

"It was the worst day of my life. I got on the phone and called Catherine back here in Virginia and said, 'You better get out here now. We might not be around very much longer.'"

They both laughed out loud, and he set his napkin on the dinner table. "We've had more than a few moments like that, haven't we?" he said, looking at her as she looked back, knowingly, from across the table.

"Remember when the *Challenger* exploded?" she said.

"It was like we had a number of pretty nimble people in the company who could move us down the field like good running backs," David said. "But the *Challenger* destroyed our ground game, which was supposed to have been TOS, which is where we expected to make the money that would build the company. All of a sudden we needed one great gazelle, a guy with big explosive ideas who could go for the bomb and change the game in the blink of an eye."

"You're talking about Antonio," Catherine said, a sound of genuine affection in her voice. "I love that man! I knew as soon as I met him, he was the one for us. He's so great. He's like the perfect father, the perfect husband—"

"He could do all the dialogue from *Dr. Strangelove* from start to finish," David added, laughing again.

Elias, a professor of astronautics and aeronautics from MIT, had not received the warmest welcome from the company's "runnning backs," who saw him as an egghead with little practical experience. But he did bring the idea of an air-launched rocket, and that catapulted Orbital into the marketplace.

"Antonio won the genius award," David said, referring to the inscription at the Wright Brothers' monument at Kitty Hawk, "Conceived by Genius." "Bruce Ferguson won the inspiration award, but Antonio was our genius."

The conversation entered a lull, and David offered to open a second bottle of wine. "Here," he said, "look at the label."

A California brand. An unfamiliar Chardonnay with a winged horse as its emblem. Pegasus.

"Scott bought these for us," David said as he poured.

After dinner I kept David engaged with questions not about the corporation's past but about his strategy for the future. He humored me, and I stayed too late.

At one point he asked if I had seen a column that had appeared recently in the business pages of *The Washington Post,* announcing the death of the Space Age. A similar piece had been published as a cover story in the newspaper's Sunday magazine. Naturally, I had read them.

"What did you think?" I asked—my own opinions hadn't taken shape.

David shrugged his shoulders. "Maybe we are at the end of an era, who knows?"

"It wouldn't surprise me," Catherine said. "They keep sending these astronauts up into space. . . . After a while you wonder what they're spending all this taxpayers' money for. They always say something like, 'Oh, we're studying the effects of weightlessness on the human body,' and you think, 'Hey, I thought you already did that.'"

But it wasn't so much the efforts of manned space missions that bothered David as pork-barrel politics that standardized the food chain of government subsidies linking Washington to a network of political districts around the country, populous places where the aerospace community could be counted on to influence congressional elections.

"Maybe the Space Age is coming to an end," he said, "and maybe it doesn't deserve to have a second life, at least as we've known it."

At eleven o'clock, when I finally picked up my jacket to leave, David excused himself to make a phone call. A team of Orbital engineers had an important suborbital launch scheduled the next day, and he wanted to call California to make sure it was still on track. He would get up the next morning at five o'clock and drive to work so he could tune in on the countdown.

Catherine, I think, had grown weary of the conversation anyway—always aerospace.

On my way out I noticed a framed illustration of the winged horse, a stamp from the wine label hung like a poster on the wall. It seemed to me suddenly that all around the small house, David and Catherine Thompson had secured emblems of a mythology that remained strangely vital to their life, like Pegasus and the Wonder Boy neon sculpture, pieces that still represented the origins of their original and most significant dream. The images were odd, though, like cave drawings reflecting a life that once was lively but whose meanings were now buried and unrecoverable. Except for those men and women whose youthful exuberance they still expressed, those images now belonged to another time.

Would they be meaningless in the next stage of corporate life?

Who could tell?

"Yes," Catherine said, as I stood there gazing at the Pegasus picture. "I guess we have a few reminders of that horse around here. You know, Scott made everybody T-shirts, too."

But that was many lifetimes ago—when they were just kids beginning an awesome and heroic mission.

CHAPTER ELEVEN

July 1993

Bet the Company

On July 1, deadlines passed without anyone noticing that the d
marked the anniversary of the satellite's original launch date. Stran
games swamped the executive staff: the memorandum of understandi
between Teleglobe and Orbital terminated and private, tempestuous nego
ations reached a crescendo in David Thompson's office. Distracted by un
solved technical problems, the satellite engineers barely recognized what d
it was, much less recalled the anniversary of their first big slip.

Expecting again to launch within a year, the engineers continued
exhaust themselves like missionaries intoxicated by high ideals. They go
siped about that class of project managers and executives they called "T
Elephants," and wondered how many new contracts and lucrative handsha
deals were unfolding in the company's rapidly growing corporate offic
Rumors came and went about Orbital swallowing up other more traditio
corporations, such as Ball Aerospace and Fairchild Space and Defense Corpo
ation, good rumors mostly that buoyed their expectations of a brilliant futu

In fact, throughout the industry, news and rumors about Orbital's n
youth culture still created a stir. David Thompson's face appeared in ads f
aerospace conferences around the country. He became a sought-after speak
the one true visionary of entrepreneurial Microspace. Publicity about t
company's newest launch vehicle, a larger version of the Pegasus, call
Taurus, cropped up often in the mainstream press and trade journals.

Despite setbacks from a few errant launches, the company had passed si
nificant milestones during 1992. By the end of the year, ten years after t

company opened for business, Orbital had seen the successful maiden voyage of its TOS booster and started the Mars Observer on its long path from Earth. Further advances in the era of Microspace came with improvements in the company's Pegasus rocket technology and an especially encouraging backlog of launch orders and options totaling $1 billion. Orbcomm also raced into the forefront of its larger competitors by staking a spectrum allocation at the world radio conference.

The news remained sanguine through the spring of 1993, with the successful launch of the company's first commercial payload, a Brazilian satellite. Privately the company celebrated the good news that Jan King and his team at Boulder had placed two suitcase-size experimental satellites aboard the BrazilSat and sent them piggyback into low-Earth orbit, where they would gather spectrum data for Orbcomm through the end of the year. A few Wall Street brokerage firms also started to tout the company as an undervalued gem, cautioning only that the company still needed a $100 million investment to finish the first phase of the Orbcomm constellation.

Even with the loss of several valuable Star Wars defense contracts at the end of 1992 and withering prospects for defense dollars as the cold war ended, none of the engineers suspected that dismal trends in the rest of the industry would ever apply to them. Although other multinational investors failed to line up to bid against Teleglobe for Orbcomm, and although David Thompson announced widespread cuts in insurance coverage and other benefits during the first half of 1993, no one in the rank-and-file took seriously occasional rumors about a coming plague of financial turmoil. The company's combination of entrepreneurial zeal and the novelty of its task—a race into orbit against every other lumbering aerospace behemoth—kept them in a state of denial.

At the Elephant level, on the other hand, lost revenues from NASA, the air force, and DOD continued to feed a climate of distress. During "Red Book" management meetings, where executives conducted weekly top-level briefings, the old joke about aerospace—"The closest thing to perpetual youth is a government contract"—couldn't raise a smile. Ironically, the young company that had boasted that it would lead the industry in opening commercial space markets found its very existence threatened because of the sudden loss of a regular stream of government contracts.

During the spring, when Thompson announced that all employees would begin paying a weekly fee for insurance, cover their own dues for annual professional memberships, and shell out fifty cents for a soda, Orbital's engineers reacted with utter disbelief.

In the lab Gurney and Denton complained endlessly about rising insurance costs. Dobbles fired off a tart letter to the company's business manager.

Krebs drew up a list of merciless observations and sent them directly to David Thompson. The loss of free soft drinks, in particular, suggested to them, not a financial threat, but a dramatic, calculated change in corporate culture. After all, employees at Apple Computer got free soft drinks. All the great start-ups supplied their troops with a stash of Cokes. The protest grew so fervent that David Thompson eventually compromised by installing soda machines that required only 25 cents a can.

While the chief executive acted to curtail the company's first seismic wave of dissent, Bruce Ferguson, the lead financial officer, exhaustively reviewed the company's fiscal performance and judged its future against industry trends.

One day he came across an article in the day's interoffice mail. In an issue of the *Aerospace Industries Association* newsletter, he found a summary of the net loss of jobs and revenues in aerospace since 1990. Missile and space-related manufacturing jobs had declined in 1992 for the fifth straight year, dropping 134,000 positions. The total number of jobs in aerospace had declined to its lowest level in ten years, keeping pace with a steady downward trend of 9.2 percent, 9.7 percent, and 13.5 percent each consecutive year since 1990. A further loss of 130,000 space-related jobs was expected by the end of 1993, and the dramatic slump in employment—while wages increased steadily—showed no signs of leveling out.

In the margins Bruce penned a note, underlined it three times, tagged the page, and routed it directly to David Thompson.

"The entire industry is in free fall!" it said.

The engineers didn't know it, but on that first day of July, Orbital was teetering on the verge of layoffs and a dangerous descent.

Steffy's engineers came in early that morning. They fed delicate crackerlike electronic boards into the maw of a thermal chamber, arranged pencil-size torque rods onto the pedestal of a rocket-force vibration table, and laid their most precious sensors atop a steel beam for assault by the thunderous whacks of a 16-pound sledgehammer. The torture of satellite testing continued in the high bay while the gamely dealings of Teleglobe went on unnoticed in the executive suite.

A half-dozen men and women in white suits covered their hair with cloth napkins, wrapped their shoes, and walked into a room made of what looked like a gigantic transparent shower curtain. Inside the clean room, a team of engineers wrapped various components on another new satellite—Apex, the Advanced Photovoltaic Experiment satellite—in thermal blankets.

Outside the clean room Mike Dobbles watched, feeling a breeze from the room's circulators puffing cool air around his ankles.

Dobbles had completed his first cut for the propulsion system. Surprisingly, his success and reputation for pluck and persistence had attracted attention from other project managers. The Freshout was suddenly in demand as an engineer.

Over the past year Dobbles had experienced almost a religious conversion to the Orbital way of doing things. As he grew in technical knowhow, he had also conducted background research on the company, ordering Harvard case studies, gathering viewpoints from industry analysts, and even seeking the expertise of the company's own accountants to give him a picture of the corporate history and explain in detail how the business grew.

He was awestruck. Orbital's corporate strategy was inspiring, something brilliantly conceived, clearly on the path toward execution. In few places could a lowly engineer call the chief operating officer, as he had, and hire on as an interpreter during a two-week investigation of satellite markets in Russia. Nowhere that he knew could a beginner take a hand in three or four equally exciting space projects at one time, or stand alongside fewer than a dozen engineers in a lab to design, build, test, and launch a system of global communications potentially worth billions of dollars. The technology was so small and nifty, the corporate structure so flexible and creative, the master plan so clever and commendable that every day for him was like a feast.

As a young man who learned best by listening well, Dobbles also soaked up technical knowledge quietly and voraciously. Taking in conversations in the labs, he sometimes discovered the clues he needed to improve his own work being bandied across a lab bench. When he couldn't resolve a problem on his own, he never hesitated to search for an expert—inside or outside the company. He even paid one of the technicians to teach him the art of welding and took a course at a local college so he could learn more about digital electronics. Within a year of starting his first job out of school, Dobbles's trajectory along the learning curve had risen like a kite.

Outside the building Mark Krebs settled his dog in scant shade beneath a hedgerow. When they had arrived at eight A.M., waves of heat rose steadily from the black pavement of the parking lot and came swelling off the brown brick Sullyfield offices. Krebs tried to sneak the dog inside, but they were discovered and the dog was tossed out before noon. The only option now was to settle the mutt by the bushes and pour buckets of water on her periodically during the day.

Enduring his first summer in muggy northern Virginia, Krebs had been awake since four A.M., when he found himself lying naked in bed with his wife with the windows open while a rattling fan blew warm air into their tiny apartment. He had consumed a gallon of ice water during the evening and taken cold showers periodically during the night. He passed the morning weighing several dilemmas: He had not sold his home in L.A.; he was on the verge of bankruptcy; his wife wanted to move back to California; his control system still did not meet spec.

The most recent addition to the spacecraft, an Earth sensor, was probably the last device Krebs could add to help the satellite find equilibrium in its delicate orbital ballet. The only other devices in the subsystem included torque rods, a magnetometer, and a gravity gradient (simply, a dead weight at the end of the antenna boom). Together the pieces amounted to such a wimpy and inexpensive combination of implements that Krebs referred to them collectively as the "cheap-ass system."

The cheap-ass system had no moving parts. Primary control of the satellite came from the two torque rods—small magnetic bars that could be electronically pulsed to set up a dipole in the satellite responsive to the force of the Earth's magnetic field. Only with very precisely timed interactions against the Earth's magnetic forces could the satellite turn itself and maneuver in space.

The cheap-ass system also depended on the spacecraft knowing its orientation relative to the Earth and Sun. The new Earth sensor encased a delicate prism and germanium lens that would read differential infrared emissions at the Earth's horizon and use the data to help gauge the satellite's on-orbit orientation. The magnetometer would measure the direction and magnitude of the Earth's magnetic field. As sensors from the two instruments fed a continuous stream of data to the attitude-control electronics, the ARSE (attitude reference system—"the po' man's ARSE," Krebs called it) combined the data. Using the sensor-fed information in combination with data from a previously designed estimator to anticipate the known changes in direction of the magnetic field, the attitude-control system could determine how to pulse the magnetic torquers from moment to moment.

Krebs's main problem was, as is often true in engineering, the laws of physics. In this case physics told him he could only generate a torque perpendicular to the magnetic field, or into the direction of the field, but not around it. Practically speaking, this meant the magnetic torque rods could only produce control in two dimensions instead of three. Leaning against the Earth's magnetic field, the torquers could give the spacecraft authority in any combination of two dimensions but not the third, such as pitch and roll but not yaw, or yaw and pitch but not roll.

Since the satellite simultaneously needed to point the solar arrays at the sun well enough to maintain satellite power at 95 percent efficiency and aim the antenna directly at its bore site on the Earth to within 10 degrees of its center, the entire control system required an enormous—"buttload," Krebs said—amount of data processing. The system not only had to keep the satellite informed of its position continuously, but it also had to predict where the magnetic field would be at every second so it could set the torque rods and plan adjustments to take advantage of the Earth's magnetic forces well ahead of time.

These were the kinds of three-dimensional problems that invaded his dreams and kept him tumbling, like an unbalanced satellite, in bed on sweltering summer nights.

As he bounced up the steps to the second floor, he ran into Grace Chang, who at that moment was headed to his office for their first meeting of the day.

"I lose sleep over this stuff," Krebs complained as they marched down the second-floor corridor. "I don't know why. It's not a good survival technique."

"Yeah?" she said. "And every night I go home for lunch-slash-dinner and remember there was something I did that was terribly wrong."

Grace had her own problems: a contract dispute over the Earth sensor; the new sensor appeared to be far from meeting technical requirements.

They met for over an hour, an aeon in Krebsian time. They needed more detail in the present orbital simulation model; they needed someone to begin running simulations continually and finding new options for pointing the solar arrays and antenna. They also needed a study of the satellite's gravity gradient to help them decide what kind of dangling mass to add at the tip of the antenna, a passive control device whose dead weight would act like a pendulum for further stabilization.

While Krebs worked on his equations, Grace climbed up on his desk and added more detail to a three-dimensional sketch that described the geometry of their Earth sensor.

"That looks like a pumpkin, Grace," Krebs said, looking over. He had finished an equation and punched up another simulated orbit.

"It's not a pumpkin," Grace said, jumping down off the desk. "This is the area—r—and this is theta, nice and constant, and this is . . ."

She had picked up Krebs's habits and was dancing in place as she described the drawing.

"Well, I defy anyone to understand that after you're done," he said.

"But I thought this is just what you like," she said smartly. "Gratuitous complexity."

He turned back to his monitor to see how the sim was coming along. "Grace," he said, "why don't you sit down in a chair?"

"I don't know, I'm worked up. Can't you tell?"

Krebs made a few corrections to his equation and entered it again into a new sim. He turned back to Grace, now sitting cross-legged on the desk.

"Look," he said, "I made this list of things that are bothering me last night because I—"

"Couldn't sleep?"

"—couldn't sleep. So I want to ask you to look at this and talk to me about it. Or maybe I'm worrying too much and you need to tell me what's not a problem."

He handed her the paper.

"I don't want to talk about it now," he said.

They both knew their attitude-control system was on a critical path, lagging far behind everyone else on the team.

Grace jumped down from her perch and stomped her foot.

"You know, Steffy keeps saying he wants us to do testing on the Flight vehicles—"

"Flightlike configuration," Krebs corrected.

"We can't even do it in a Qual-like configuration!" Grace said. "We don't have all the hardware, for one thing. The vendor keeps screwing around—"

"I've been globally unsuccessful in getting the world to think that our shit is important," Krebs said.

Suddenly, Grace had an idea. "I know what, let's tumble the spacecraft a couple of times. Once we've launched, on the tenth orbit, we start a tumble."

"Martyr," Krebs said.

"Yeah?" she said, taking to the idea. Then looking over his shoulder at his monitor: "Sorry, Mark, your sim's going nowhere."

Krebs spun around. Sure enough. He stepped through the data. Grace folded her drawings and the list he'd given her, and started out the door.

"Come on down to the high bay when you're done," she said.

But his body was already keened into the monitor, studying a new set of equations.

"Man," he muttered, "if I could just pass one more function through this thing, I'd be fat."

Late in the morning Steve Gurney wandered into Rob Denton's office to hash out a rumor he had just heard indicating that the Teleglobe deal had gone sour. A moment later Gregg Burgess stuck his head in the door with an update.

They had both pegged Burgess as Steffy's new schedule enforcer, hired, like Krebs, out of Hughes to be the satellite's overall systems engineer: another midcareer techno-weenie with an MIT beaver ring. Most of the time they took an attitude of bemused detachment with him, but Burgess was almost impossible to shake.

"It's not as bad as it sounds," Burgess said. "Our people are just regrouping."

"I thought Teleglobe said no," Denton answered. "But is it 'no forever' or 'no for now'?"

"There's some kind of hitch," Burgess said. "Don't know if it's technical or if they just want a bigger piece of our pie. Could be they don't want to spend $80 million now, so they're smoking us out to see how much of the store we'll give away. But they're definitely not walking."

"I heard they had two problems with us: technical and schedule," Denton said. "They're saying no way we'll launch in December—it'll be next June at the earliest."

"Who? Us?" Gurney said, with his usual note of irony. "Miss a schedule? Maybe they were disappointed with our technical review."

"Why don't you call David Thompson and get the whole scoop," Denton said, waving him off.

"Oh, good idea," Gurney smirked.

"He has an open-door policy, you know," Denton said cynically.

"So," Burgess said, changing the subject, "what's your schedule look like?"

That was the real reason Burgess dropped by—as their new boss, he had to keep the Freshouts focused on deadlines.

Gurney ignored him and turned to Denton. "Do you think people would pay money to send their ashes into space? We could just tell the world we have a little extra room to piggyback a few cremated bodies with our next launch."

Obviously, they had no intention of discussing his schedule with Burgess. But Burgess was patient.

Finally, Denton relented: His work on the battery-control regulator and flight computer had been delayed by problems with the OSX operating system, which Morgan Jones had under review at that very moment in Antonio Elias's former executive office.

Burgess listened carefully and nodded. Noting that the buck had been effectively passed, he turned and left to find Morgan in the executive suite.

"So he's treating the Teleglobe rumors rather glibly," Gurney said as soon as Burgess was out of sight. "Or maybe he's just trying to boost morale."

Denton laughed. Management wannabes, they believed, always thought it was their the job to "boost morale."

"I mean, David Thompson must feel like a ton of bricks just fell on him," Gurney continued. "It's like, why didn't we have someone else lined up for the partnership? What did we do, just blow off all the other offers?"

"I don't think so, Gurney," Denton said.

"I want to just ask him, like, 'What business are you in, buddy?'" Gurney sounded serious, like he might just take his questions directly to the executive suite. "I'm sorry, but you don't negotiate with just one company. It's a bad tactic. I mean, that's not how you play the game. You don't just sit there and take one offer.'"

"Gurney," Denton said, "did it ever occur to you that they don't have any other offers?"

"If they just had one other company on the line, they could play them against Teleglobe," Gurney said.

"Hello. That's what I'm telling you. There aren't any other companies," Denton said.

Gurney didn't seem to want to hear what was being suggested. It couldn't be true—not at Orbital.

"So we've got to scrape up $80 million somewhere else?" Gurney said finally.

"You know, maybe they're just bartering: their international contacts, for more than half our business," Denton said. "I heard they want 51 percent of the partnership, but they won't make an investment until we're in operation. That way we take all the risk and they still become majority owner."

"Oh, good deal!" Gurney sneered.

"I think this is rather crushing news personally," Denton said. "I was looking forward to building the constellation."

"But if we fail—if Orbcomm fails—then Orbital is what? History?" Gurney was not really asking the question but making a statement to see how it would fly.

The conversation drifted into silence, interrupted suddenly by a call over the PA system from the receptionist desk downstairs.

"I heard that her hours have been cut back," Gurney said. "She only works from eight-thirty to five-thirty now. Remember how it used to be? We'd be in here late at night, and we'd get these calls over the loudspeaker. 'Denton, Rob Denton, your wife is on the line. . . .'"

"I don't understand why they can't pay her the extra ten bucks to stay a few more hours."

"I think it's because the Elephants moved to Steeplechase," Gurney said.

"Right, no more big dogs around here to complain about not getting their messages," Denton said.

"Left us a lot of room, though," Gurney said, stepping out to look down the hallway. Burgess was standing outside Antonio's old office, waiting to chase in after Morgan and ask to see his schedule, too.

"Yeah, how would you like to have that office?" Denton asked, peering out to see.

Gurney rolled his eyes. "Hell, yes. I'd take Bob Lovell's any day. That's a sweet space, man."

Down another hallway Dave Steffy had unpacked his briefcase and called in his associate, John Stolte, to look at yet another revised schedule.

Dutifully, they abided by David Thompson's private request to discuss layoffs, which could come by the end of the year. Steffy and Stolte made a feeble attempt to identify specific people to expunge from the payroll. Not a single name emerged. Groaning over the gruesome and depressing exercise, they turned back to their spreadsheets to calculate the next schedule slip.

In Steffy's experience Orbital had always managed, often fantastically, to swing critical financial deals at the last minute. The team of Thompson and Ferguson, in particuar, had an amazing ability to forge partnerships, sway investors, attract support for public offerings, and nail God only knew how many extensions in the company's line of credit. During his five years with the company, Steffy had watched them suck five or six large infusions of capital into the coffers. The regular occurrence of last-minute deals established a phenomenal pattern. The only difference was, this time the deal was a little more personal—Steffy needed Teleglobe's $80 million in cash to keep his own project alive.

His allegiance to Orbital, like his unrelenting optimism about the Orbcomm project itself, came at least in part from watching these life-preserving cash deals occur and then slowly enhance his own bank account. If ever a Freshout expressed a concern about whether the company would survive its Microspace infancy, Steffy had no trouble offering reassurance. He thought he had a true sense of how much of the company's promise still remained to be capitalized.

For him and his wife, Jennifer, benefits just kept accruing. Raises and bonuses made it possible the previous year for her to stay at home with their two kids and work part time. They had saved enough money to start building their own home in Great Falls, Virginia, and they even made a down payment on a vacation beach house a few hours away in Nag's Head, North Carolina.

Despite the way Teleglobe's inspectors carefully stalked Orbcomm's progress from the high bay to the board room and questioned his overly optimistic schedules, Steffy believed it was reasonably safe to assume that the Canadians' proposed $80 million investment was all but deposited in the bank.

Recalling the previous eighteen months, he would proudly point to exceptional moments of progress from his largely inexperienced team. Particularly in the past six months, Tony Robinson's brilliant new antenna design and Mark Krebs's smart redesign of the attitude-control system stood out as specific innovations that had moved the team dramatically forward. Certainly they had taken risks by not pausing to document their work—most of the satellite's design, in fact, existed only in the engineers' minds, not on paper—but they had saved large amounts of money by keeping the team down to a preposterously small size. With only about twenty engineers in all, his young engineers had created efficiencies by traveling frequently, hiring their own vendors, and conducting their own design reviews. They had also accepted the challenge of designing the Lightning Sat and GPS Met satellites on top of the work with Orbcomm.

Even though he felt increasing pressure to meet financial goals at the end of each quarter and sometimes wondered if the corporation could retain the excitement of a small company, he had faith that David Thompson and Bruce Ferguson would lead them through the roughest waters. Even on days like this one, when he was asked to consider layoffs and an accelerated schedule, he believed Orbital would succeed and do so without growing dependent on federal subsidies, pork-barrel politics, or a stultifying bureaucracy of its own making.

Among his latest instructions, Steffy had been told that his bosses expected a launch no later than February '94, a six-month target that was as absurd as the thought of layoffs. Looking at their calendars, he and Stolte decided to plan one schedule with a launch date of February 28, and then work out a second, more realistic, private schedule that they could track for themselves. The real launch date, they decided, would be July 1994—a full year away.

Turning to review the program's financial estimates, they also saw the projected cost of their two satellites had risen to $31 million—not a great surprise considering the up-and-coming need to contract three more antenna consultants, expand the team of engineers, and absorb a growing accumulation of cost overruns due to vendor workmanship problems and contractual nits caused by the elusive deadlines. As with the launch date, they noted the actual cost, then cheated a little on the figure they would report to the boss.

Instead of $31 million, they simply rounded down to a more politically correct sum of $30 million.

The art of tweaking finances and schedules was never as precise as in engineering. And the real risks never seemed as great.

Gregg Burgess stood outside the old executive suite and stared at a handwritten sign taped to it saying "OSX Cafe." Dangling from the door handle was another sign taken from a nearby hotel—"No Service Today." He paused a moment, perhaps deciding what gambit to employ, then grabbed the handle and entered the room. He slipped in with all the deliberation of a burglar and stood quietly on the perimeter for Rob or Morgan to notice him.

Morgan Jones and Rob Phillips did not look particularly crazed, even though they had spent the past ten days and nights digging through miles of mind-boggling code.

With the blinds pulled up, they could approximate the time of day, if not quite the day itself. They sipped coffee by the window and stared down into the small face of the gray monitor like two ice fisherman poised over a hole. They were fishing for bits.

From Gregg's distance, their conversation sounded polite, calm, and earnest. Looking into the logic analyzer, Morgan whispered, like a magus, "Make it so!"

Rob rose up and turned to face into the blue tint of his terminal while Morgan remained facedown over the screen of the analyzer. Rob tapped his keyboard, and Morgan studied a flash of bits blazing across his screen.

The executive suite's air-conditioning gently brushed through their hair. A late-morning splash of sunlight warmed their backs and shoulders as they continued sending and documenting the passage of bits, still ignoring Gregg, who stood in the doorway, just within their field of vision.

"So how long will it take you two guys this time?" Gregg snapped. He had waited too long for recognition.

Rob slowly reached for a white bag and offered a bran muffin to Morgan. Then he turned to Gregg.

"Be quiet! Can't you see we're designing?"

"Would you like to let me in on more than 'We are designing'?" Gregg said. He bit into a carrot, his customary morning snack.

"No," Morgan said.

Rob and Morgan laughed and turned back to their work.

"Okay, gentlemen," Gregg said. "What are you doing?"

"Ah," Morgan said, "the true manager comes out!"

"His true colors!" Rob snickered.

"Exactly," Morgan said. "In other words, I should either tell him our schedule or run like hell. Isn't that what you're saying, Gregg?"

Gregg had come to expect their exchanges to be prickly but not quite so rude. He took another bite of the carrot.

"Okay," Morgan said, "it's like this: OSX isn't doing anything it's not supposed to be doing. We've convinced ourselves of that. We've squeezed it as much as we can, but we can only clock out four or five microseconds here and there. It's not enough."

"You've whittled it down to the nub?" Gregg said.

"Unfortunately," Morgan said.

Rob laughed. "The next step is discovering how to transfer infinite amounts of data with no overhead and no time."

"Zero time," Morgan added.

"This is a riot!" Rob said, looking at his friend. "I love this. Zero time: If we can do that, we're home free—zero zero zero zero."

Rob traced a finger across the analyzer screen while Gregg knit his brow and stepped over to read the data. These two continued to challenge his patience.

"If I were a cynic," Morgan said, "I would say we're falling in line with management schedules."

"What does that mean?" Gregg said, trying not to sound defensive.

"Negative time," Morgan said, laughing again. "Our goal is to create negative time. Problem is we're just approaching zero."

"Given that we are only mortal," Rob said.

"Yes," Morgan agreed.

They sounded as if they had hatched from the same egg. An acerbic tone, clipped diction, a decidedly indifferent air—shared traits that either had merged during these weeks together or evolved out of their combined chore of scouring minutiae for clues to the satellites' most confounding mystery. They had clocked and recorded the passage of thousands of bytes, reviewed them, filed them away, and studied the paper trail. Like mathematicians engaged in theoretical research, they discussed evanescent time, microseconds, nanoseconds, the very breath of time, as the analyzer pictured the frozen blip of every bit's passage through the satellite's six 68302 processors.

For the last several days—from Saturday to Tuesday—he and Morgan had categorized to the microsecond the time OSX expended to service an interrupt, an interrupt being a break in the computer's regular routine to act on a new message and then, once done, resume the routine. For instance, the

Motorola 68302 processor would take a message sent from the ground, stuff it into a buffer, and pass it along to a digital signal processor (DSP), which would prepare the message for return to another site on the ground. It was a simple maneuver, at first glance, since the message simply needed to pass from site to site on the Earth, using the satellite as nothing more than simple repeater, a very high antenna revolving a few hundred miles above the ground.

The timing requirements posed a significant challenge, however. Through a tightly scripted plan, data from a gateway Earth station would pass to one of twenty-six Orbcomm satellites at a rate of 57.6 kilobits per second, and the data would be returned at the same rate. To keep from mangling messages that might arrive simultaneously either at the satellite or on the ground, a timing sequence for passing data had to be established before the first bit was transferred. Every millisecond mattered. Each second had to be perfectly synchronized between the earth system and the satellites, giving every message a unique starting point to make the link.

On board the satellite the schedule for converting data from an analogue to a digital format and for preparing its return to Earth required the same tightly bounded time restraints. Alerted from the ground that data were coming, the satellite would prepare a unique frame in which to pack the data. Once received by the satellite, the data would be squeezed into the frame and transferred back to the ground at the moment the ground station requested it. Only an exquisitely timed synchronization scheme would prevent the constant flow of messages from entangling and crashing the system.

Annette Mirantes originally discovered the problems during her tests in the lab. She and Gurney had experienced trouble with OSX even before the various subsystems were built, but once software code from other parts of the satellite came into play, she began to chart ways that specific features of OSX repeatedly clogged the system as it answered health and maintenance calls and serviced other kinds of interrupts that were naturally part of the operating system's regular tasks. Annette found that the operating system actually took on too much work. OSX, she concluded, had been written with so much ornament that its intention to juggle many tasks simultaneously made it almost impossible to thread messages through the satellite without leaving a wreckage of bit packets strewn in its wake.

The problem with OSX was that it was neither as fast as everyone expected nor as efficient as was necessary. The operating system created a large number of "housecleaning" tasks, checked for errors and logged them, distributed a regular "heartbeat" pulse that synchronized the timing among the various subsystems, distributed the token among boxes for intersatellite communication, collected telemetry, buffered messages, and gathered information collected in the buffers. It handled tasks that in previous satellites would have been conducted by hardware only and provided some special sophisticated features that saved considerable weight in hardware but also slowed the satellite's primary task—passing messages from Orbys or Earth stations on the ground—to a deadly slow pace.

For messages to pass unimpeded from a gateway Earth station or a subscriber terminal from the ground to the satellite and back again to the ground, OSX had to be extremely nimble in caring for the internal needs of the satellite itself. Unfortunately, the software still was not fast enough.

"Hosed" was the word Annette used when she originally reported her findings.

To pinpoint what hosed the system, Rob and Morgan had holed up in Antonio's old office and spent days merely tracing the passage of 435-byte messages in a simulated exercise, from the moment they entered the satellite at the receiver to the time they exited the gateway transmitter's DSP.

Gregg Burgess should have known the engineers would try to obfuscate by swamping him with details as soon as he entered the lab. After all, Morgan and Rob would never admit defeat.

But Gregg was too cagey to be caught in that muck.

"It sounds like you two are on top of things," Gregg said at last, after hearing a long explanation about why they could not be held to a deadline. "So why am I not immensely relieved?"

"You will learn eventually not to ask questions you don't want answers to," Morgan replied.

"Only because you have made the classic mistake of creating more questions than you have answers to," Gregg said.

Rob sniffed, "We have gathered all the data!"

"And," Morgan continued, "we're now going to do all the things we set out to do last Sunday afternoon on *Thursday* afternoon. So put that in your pipe and smoke it!"

"Well," Gregg said, still unflappable, "then this is good. You have a real job to do now. The only question is, how little service can OSX provide and still get away with it so the processors can do what they need to do."

"And the further question: How much time can OSX spend housecleaning, and how much time can it spend doing real work?" Morgan said.

Gregg picked up a blue marker and stepped over to the white board to draw a message-passing scheme—the 435-byte message and then a series of interrupts for clock ticks and tokens.

"Just suppose," he said, "the second and third interrupts come at the same time. You can't get the token back until you service the second interrupt . . . and if you get several messages in a row and you don't have the token, you will never get out of the loop. It's endless."

Morgan looked startled; Rob, perplexed.

"Daaaah . . ." Morgan said.

"We have a problem," Rob mumbled.

"It's not looking very robust, my friends," Gregg said happily.

Rob turned to Morgan and whispered, "We should have just thrown him out when he first came in here."

Gregg smiled, a haughty expression. He said: "Keep me posted, okay?"

"You!" Morgan said, pointing a finger at Gregg. "Recompile thyself!"

Gregg walked out, a smile of satisfaction still on his lips.

Morgan looked at Rob, still perplexed.

"Duh," Morgan said.

"Where were we?" Rob said.

They stared into their monitors for a moment and then looked away.

"You know, there's a new coffee shop in Centerville," Morgan said.

"Good point," Rob answered.

The two young men stood up in the sunlight, stretched, grabbed their notebooks, and left the executive office talking about creating a happy face for the code, something for the ground station operators, "a figure of merit," Rob said. "Morgan is happy."

"With the edges of the smile coming down as the satellite tumbles?" Morgan said.

"Exactly," Rob replied.

But they were not smiling as they left the room. Suddenly, they needed some kind of hot, dark brew—anything to wrench them wide awake and pitch the two of them into another afternoon of flipping bits.

In the life cycle of any successful entrepreneurial company, long after the IPO and the emergence of the corporation as a valid competitor with multiple product lines, way beyond that moment when investors switch from being gamblers to being mere speculators, the transition to maturity comes with a jolt.

The company's profits and cash flow meet most capital requirements, but rapid growth rates pose a serious and immediate challenge. The entrepreneur is forced to think more about the overall corporate direction, extending product lines, employee morale, internal communications, and long-range planning. In time, as potential merger or acquisition candidates come forward, the entrepreneur may even fade as the central figure in the company, giving way to an amorphous collection of lawyers, accountants, executive committees, marketing experts, and business managers who have stepped in to create a more complete institution out of many fractured parts.

Questions about survival are still asked with urgency, as in the early days. But so far removed from the lab or program manager's concerns, a rallying cry may not sound the same note of warning as when the entire company once shared news over a pizza lunch inside the chief executive's conference room.

If a growing high-tech company has the traits of a romantic mission, its maturing self, in its resolve to preserve its character, to satisfy its investors, and to compete among a higher order of corporations in its field, will shift sights from technology innovation to the more conservative art of corporate husbandry. The entrepreneurial figurehead may sell out and start another new venture. Young Turks in the labs will wonder what happened to skew the original vision. The call for sacrifice still comes but, they may ask, for whose benefit?

The original clarity of a single goal and shared sense of purpose sometimes recedes behind a dense, continuous discussion of production rates, office politics, margins, and quarterly reports, all of which soon become as essential to the ongoing life of the corporation as either the products themselves or the flash of genius from which the essential momentum surged. The quest for great challenges and the company's once-thrilling esprit de corps are recalled, but only as they fade into the realm of lore.

Everyone at Orbital saw indications of corporate maturation, but until Teleglobe entered the picture, not everyone believed the changes would ever occur so rapidly or with such consequence. Particularly given the difficulty of negotiating a partnership with the Canadians, which eclipsed the thorniest problems Orbital's executives ever experienced at the bargaining table, it was hard for anyone to see the corporate changes so swiftly under way.

The Canadians' negotiating tactics had bewildered the executive staff. Indications that they would no longer bargain would precede a series of letters from Montreal laying the groundwork for a host of negotiable concerns. The good old standard route to deal-making (pitch-offer-negotiation-contract) never had a chance with a telecommunications giant apparently accustomed to leading the process like a tango across a dark room.

No one at Orbital expected it to be easy, but neither did they anticipate such ambiguous responses. Preliminary signs in the spring of 1993 pointed to a quick consummation. For instance, after Teleglobe's consultants listened to a series of detailed technical and financial reviews at Orbital, the Canadian attorneys worked with Orbital's legal staff to hammer out an apparent deal in May. But just as David Thompson and Bruce Ferguson awaited a final document to be signed, the Canadians withdrew, blaming some vague miscommunication with their banker, who had not fully assessed the numbers.

Then the next month, in June, Teleglobe's executives sent a letter suggesting they believed the Orbcomm system would never work, but at the same time they indicated that their attorneys wanted to help Orbcomm speed along its licensing with the FCC. No one, not even Thompson, knew what to make of the continually contradictory signs.

Finally, on the last day of June and during the early hours of July 1, David Thompson held the longest phone conversation of his life—a marathon conference call that began at six-thirty P.M. and lasted more than five hours, ending just fifteen minutes before the memorandum of understanding timed out at midnight.

For the first time in his life, Thompson was forced to make a deal he did not like.

Rather than invest $80 million in the Orbcomm project right away, as everyone expected, Teleglobe's executives made a counteroffer wrapped in a surprise package that created a set of clever impediments—a gated pathway—to the money.

The proposed contract reduced the Canadians' risks considerably—and forced Orbital into a corner. The deal was this: Teleglobe would pay $2 million at once. Eight million more would follow in October, but only if Dave Steffy's engineers stabilized the satellite design and David Schoen's engineers at Orb Inc. completed a nationwide testing program to prove that interference from ground sources, such as telephone lines and trucks on the highway, would not degrade communications with the handheld Orbys. The remaining $70 million would complete the first round of funding and lead to a partnership, but only after the first two Orbcomm satellite prototypes, FM-1 and FM-2, were successfully launched and passed all of their functional tests within ninety days of nailing their orbits.

Essentially, until a successful launch and an on-orbit testing program, Orbital Sciences would still commit most of its own cash resources—more than $30 million—and bear every significant risk. In exchange, Teleglobe and Orbital would create a set of separate companies under the jurisdiction of one parent company called Orbcomm Global, in which both corporations shared exactly fifty-fifty ownership.

Lacking Teleglobe's full and immediate financial support, David Thompson saw his ambitions drastically assaulted. His moment hadn't passed, but it had certainly slipped. Instead of boasting the world's first full-time global/mobile personal information constellation—the Net lifted skyward—Orbcomm could possibly miss its opportunity and pass on as a two-satellite afterthought, an insignificant also-ran in the twenty-first-century race to the LEO frontier.

In the course of the five-hour call, Thompson weighed only three options: He could allow the MOU to expire and spend his days searching for another international partner; he could renew the uncertain task of bartering among a new list of lesser corporate investors and try to build a consortium; or he could accept the risk and bet his company on the gamble with Teleglobe.

On that Thursday morning, as Dave Steffy's team continued to struggle with the first development models, Orbital deposited a check from Teleglobe for only $2 million into its accounts. Beginning that day, the Canadians could expect to hold the Orbital Sciences Corporation to a short leash for at least the next eleven months.

No one knew it yet, but David Thompson had just bet the company.

CHAPTER TWELVE

August 1993

Snafu: Situation Normal . . .

Early in August, on a mountaintop halfway between Buffalo and Elmira, New York, Stan Ballas walked along the periphery of a cow pasture in deep concentration, studying a muddy plot of former grazing land where he would, by day's end, erect the first gateway Earth station.

Stan stepped over to a rusted trailer perched on the north end of the property, where he had withstood the previous week's rains, and walked inside to grab his cell phone. A moment later he was out in the mud again, plugging an adapter into the cigarette lighter on his Buick.

A fellow in a workman's uniform waved from the distant cornfield.

"How you doin'?" the fellow shouted.

It was the serviceman from the local utility company, just who he'd been trying to call.

"Worse 'n terrible," Stan replied gruffly.

The crystal on Stan's watch was shattered, his blue Adidas tennis shoes were caked with brown clay.

Fifty feet away in a dug-out, muddy clearing, a silver-haired man dressed all in white and a younger, scruffier man with a dark beard, named Bones, stepped off the platform of a cherry picker onto the top of a large, gleaming white ball, the satellite antenna's radome, and waited for a crane to swing a lustrous round fiberglass panel, about five feet in diameter, drifting precariously through the air, in their direction. They reached out, steadied the panel, and positioned it as a cap, like the top of a teapot, over the dome.

"Looky here," Stan said, spitting into his cell phone, "we got a 80-ton crane comin' in tomorrow, and it was s'posed to hit Tuesday, but the good State of New York wouldn't give us a permit to run, and then the good State of Virginia closed us down, and we finally got 'em both to let us back on the highway just last night. So that big boy's gonna be here sometime tomorrow morning, I s'pect, and . . ."

Stan listened a moment and kicked at a cake of clay around his Buick's front tire.

"I'm telling you it's a legal load, but it's a *wide* legal load, so I can run across it. . . . No, we don't know nothin' 'bout no bridge. . . ."

He watched Bones spin the white cap on the Earth station dome until it was level.

"Yeah, it'll stop traffic, that's for sure. Plus, it's s'posed to rain all night. . . ."

In a moment Stan got off the phone, muttering to himself, "All these detours." and spit on the ground.

Far from the swell of headaches back in northern Virginia, Stan made his part of the Orbcomm drill look relatively easy. With no more than six men, including the crane operator hired in from Buffalo, he would complete construction on the company's first Earth station: a hub for Orbcomm communications that would cover the passage of satellites across the entire northeastern quadrant of the United States.

The sun-baked mud didn't bother him. He guessed the crew would be finished by six o'clock, and by seven they'd be back at Josie's bar celebrating in the little town of Arcade, New York, tossing back beers and shooting pool.

At the moment, however, he had no electricity. An underground conduit laid by the power company just the week before apparently had snapped or severed.

Stan left his car and tromped down the hill to where the local utility man stood over an open hole in the ground.

"You're too damn slow," Stan said, with no indication that he was joking.

The utility man told Stan he was late because he couldn't get his truck up the hill through the mud.

"You guys were s'posed to run us some power in from that box down there," Stan said. He pointed down the path to a fence where a few sloe-eyed cows stood. "I don't know if the conduit's crushed or broken, but I 'spect we gotta dig it up."

The utility man looked blankly at Stan and said nothing.

"It could be water's got in there," Stan said, "and the weight collapsed it. I 'spect if we dig up 200 feet from down at the gate to right here, we might just find the trouble."

The utility man still didn't say anything, merely looked back at the hole Stan had dug himself that morning, exposing a five-foot section of conduit.

Stan picked up a hacksaw he'd left lying on the ground and stepped down into the hole.

"Here," he said.

With a brisk stroke across the top of the white plastic pipe, he opened a gash. A spout of water came squirting out, soaking Stan's shirt, drizzling the utility man's pant leg.

"There you go," Stan said. "Thing's full of water. Get your truck up here, and you can pump her out."

Stan scrambled out of the hole and left the man agog as he marched off to call E. J. Freyburger & Sons over in Java Village. Now that the day promised to be dry and sunny, he wanted a 'dozer to finish leveling off the ground. "I don't want the next rain to muddy up my pretty new Earth station," he told Freyburger.

Chainsaws whined around the periphery of the mountaintop. Once in a while a pine tree fell. One requirement for the Orbcomm system stated that the Earth stations' big cone antennas had to be able to send and receive messages when the satellites were as low as five degrees above the horizon. For days treetops peeking up over the mountaintop's grazing grounds had come tumbling down, exposing another patch of sky and a clearer line of sight.

For more than a year, Stan had searched the country for ideal spots to place four Earth stations. He had purchased land in the states of Washington, Arizona, Georgia, and New York, representing the four quadrants of the continental United States. Unfortunately, Stan's work had been interrupted by the Teleglobe negotiations, and no one was quite sure anymore whether the company could yet afford to build more than one Earth station. Even though Alan Parker was buying them from Orbital's own Space Data Division in Tempe, Arizona, the Space Data group had, like Dave Steffy's Space Systems group, kept escalating their costs on the project—partly, Parker suspected, to cover losses in other parts of its division. Consequently, Parker could no longer afford four Earth stations plus a spare for each. At the moment Orbcomm could afford only two, one in New York and one in Arizona. The $7.5 million tag for two more was, Parker said, blatant robbery.

But that was not Stan's concern. His mind was clearly on the day.

Under Stan everything ran on a minute-by-minute schedule. His direct, no-nonsense manner had helped him create good relationships with local contractors. His offers to hire local labor also gave him a leg up when he needed immediate assistance, as he did that morning from the utility company. He had worked to make friends during his trips to Arcade, when he came to identify the property, purchase the land from a farmer, and speak at public hearings.

He was in the good graces of the Freybergers and the local utility company. A man of his word, Stan also hired a local contractor to pour the concrete. He got his crane operator from Buffalo and called on extra hands, when he needed them, from around town. He ate his meals in local restaurants, drank beer at the town bar and grill, and had a habit of remembering at least the first name of anybody he met. He felt right at home.

Until the conduit broke, Stan thought he had settled all the outstanding issues. Phone lines linked to the power station would carry data coming into and going out of the gateway antenna either directly from the satellites or from the network control system back at Orbcomm headquarters. The power would feed a humidity-free, computerized operations room inside a concrete bunker next to the dome, which would control the Earth station virtually without any human assistance at the site.

An agreement with the Sprint telecommunications network made the carrier responsible for routing data between the Earth stations and the network control center in Virginia. It was not too complicated.

Stan had also hired the six-man team from Space Data to build the conical gateway antenna, and a four-man team, of which Bones was a member, from Ratek, a small start-up company outside Tempe. The crew from Ratek had bragged that they could finish the construction job in a week's time.

Constructing the radome, today's big event, brought all the excitement of an old-fashioned barn-raising. With major parts of the station set out on the ground like pieces of a prefab home, the two teams had started full bore a little after seven A.M. and by midafternoon appeared close to piecing everything together.

The company's president, the man dressed all in white, was presently perched atop the dome, looking a little like the main character in *James and the Giant Peach* or a Lilliputian plopped down on a toadstool. He expected to ride the coattails of Orbital's success. If the assembly went well, he would probably win a contract to construct twenty-seven more gateways for Orbcomm around the world. Needless to say, his team worked up a good sweat.

By two o'clock in the afternoon, it was so hot that most people had stripped off their shirts. Rock 'n' roll echoed inside the radome from Bones's boom box. They had hooked the white dome to steel cables that threaded down the arm of the 40-ton crane. According to the plan, the crane would lift the great dome straight up about 40 feet off the ground, turn 90 degrees, and gently lower it over the 25-foot cone antenna, standing now open-mouthed to a sparkling blue sky.

Stan handed out bottles of orange Gatorade. He said it seemed like a long time since they had all sat down at six that morning to eat breakfast at the Arcade Putt-Putt.

Around five o'clock, Stan whistled to the crane operator to tighten the cables and begin the lift. The telescoping arm on the crane climbed slowly skyward until, at about 40 feet, it tightened up on the cable, which held a ring of six straps laced into the sides of the dome. Up it went. The dome seemed to float off the ground.

One of the fellows from Space Data kept burnishing the white antenna, removing the last flecks of mud left from the previous day's rain. The antenna looked like a giant bullhorn aimed at the sky.

For miles around, homes and farms stood out on mountainsides in bucolic order. The sight of microwave towers set 25 miles apart on a distant mountain range was the only indication that a high-tech experiment had ever touched the region. A few gaping satellite dishes aimed at distant geosynchronous birds to snag television channels appeared in the trailer park off Main Street and atop Josie's Motel. But nothing like this. Nothing remotely like this.

"When I was in the air force, I built Earth stations," said Bruce Barkley, the youngster from Space Data obsessively cleaning the antenna. "But what you see happening here would never happen in the military. You could never field a small team like this. Those of us who designed it will test it here next week; the guys who built the molds are over there helping lift it off the ground.

"In the service you'd never do it like this. In the air force I'd take someone else's documentation and someone else's electronics and follow the instructions to put it together—part A to part B—and no one really gave a shit. I spent four years working on a station like this for the air force. I quit because it never got finished and the technology was already outmoded. Even now, ten years after I started, you know that Earth station's still not built? But we'll finish the first phase of this one before dark. That's what happens in the commercial world when you put your heart into your work."

A moment later the radome floated down until it hung directly over the cone-shaped antenna, gingerly balanced with guy wires held by the Space Data crew. It hovered there while the crane operator awaited instructions to lower it further.

Stan stood at a distance, knee deep in an alfalfa field nearby, snapping pictures. He kept saying the dome looked like a plump scoop of vanilla ice cream being set in its cone. He motioned to the crane operator and then watched the dome descend until it gently touched the base of the concrete bunker.

While everyone bolted the radome in place and finished off the work, Stan went back to his Buick and picked up his cell phone.

"Well, who the heck do you think this is?" he said, teasing David Schoen, who was at work in the new network control center at Steeplechase. "Oh yeah! Oh yeah! Just like kindergarten in the sandbox.... Lots of fun.... Just wanted you to know. Ciao!"

At six-thirty the sky was just turning pink. Stan invited the crane operator and the utility man to join them at Josie's for a beer and pizza.

By eight o'clock Stan and his mates sat at the bar among two long tables clanking with dozens of empty beer bottles. A few of his guys had started shooting pool and chatting it up with black-leather bikers.

From the window where he sat, Stan looked up and saw in the distance the moon as white as milk, and down below it, on a cleared-off section of grazing land, he thought he saw their white dome. Holding up a half-empty bottle of beer, he proposed a toast to his noisy crews.

"We're only a month behind schedule!" he said. "And after Labor Day, buddy, we gonna do it in Arizona!"

At last the first piece of hardware in the Orbcomm system was completed.

Alan Parker raised his tall frame off the conference table, laughing at his colleagues who complained that the cholesterol count at Orbcomm Inc. kept going up and up. The brass clips on his fatigue-green suspenders glinted against sunlight where it broke through his third-floor conference room window. His eyes sparkled.

"You cannot control it with either a change in diet or exercise," said Martin Deckett, Orbcomm's international marketing manager. He had just returned from an exhausting business trip to Montreal. "It's stress."

"Stress," agreed Bill Fox, the company's business manager, who had predicted disaster without Teleglobe's full commitment.

Alan was only amused. "Sure it's stress," he said. "But I wouldn't trade it for anything. I mean, this is fun! C'mon guys, this is an addiction!"

"An addiction," repeated chief engineer Paul Locke gloomily. Paul had just returned from his doctor with news that his LDL count had escalated close to 300. "You don't know what excitement is until you've waited for one of these launches. All you can do is sit there knowing you won't have a job in the morning if your satellite doesn't nail the orbit tonight."

Alan clapped his hands. "Whoo boy! Life or death!"

Alan seemed invigorated by the idea. In fact, ever since the Teleglobe deal squeaked through, he had been a tornado of energy, talkative and high-spirited. The July agreement certainly put the business in jeopardy, but in the past six weeks, Alan proved far more resilient and optimistic than anyone in his position had a right to be.

Teleglobe's conditional commitments had devastated his business plan. Without a major up-front investment, Orbcomm's marketing team saw prime contracts slip out of reach. The delay in funding shook the confidence of several potential clients, such as Westinghouse, and threatened to erode Orbcomm's early advantage over competitors in the fight to be first to market.

Alan gave daily pep talks to the marketing team, spoke reassuringly to the engineers, strutted around the office almost buoyantly. It wasn't that he refused to recognize the risks. One day he even called his staff together and calmly suggested that he would gladly accept the resignation of anyone who lacked the guts to ride out the storm. But he never showed any sign of doubt. Alan simply did not give in.

So he joked and laughed. Even as they started their morning staff meeting sharing data points from the staff's rising cholesterol count and segued, quite naturally, into plans for cutting back the business from a full thirty-four-satellite constellation to a slight two-satellite system, Alan kept teasing.

"Just don't take the medication if it affects your sex life, Paul," Alan advised as Orbital and Teleglobe executives entered the conference room to begin the next grim discussion. "I'd rather die of heart disease than be impotent."

Bruce Ferguson's face was ashen when he walked into the room. Teleglobe's marketing representatives, who now sat in on most of their meetings, stood politely and shook his hand. Like a man who had been forced to the brink one time too often, Bruce appeared to have aged in the past year and a half. His shirt was wrinkled. His dark hair curled up around his ears and collar, sorely in need of a trim.

"Sorry I'm late," he said. "There's been a little snafu."

"Snafu?" Alan said.

"SNAFU: Situation Normal—All Fucked Up," Bruce answered. He sat down without further explaination.

In recent months Bruce had shown greater interest in the technical aspects of the Orbcomm system. One day, for instance, he held Orbcomm's staff captive to a three-hour session designing a QWERTY keyboard for the handheld Orby messaging terminal. While Bruce acknowledged the importance of the business's industrial applications, his abiding passion was the Dick Tracy–like Orby, the device he expected would enter the marketplace as the world's first personal messaging satellite communicator. It did not matter to him if they had thirty-four satellites or only two—Bruce wanted the personal messaging application. It had always been his personal desire to see Orbcomm become the world's best search-and-rescue system, to save lives as well as make money. To him, the fullfilment of altruistic motives would be Orbcomm's first home run.

He sat quietly through the presentations, as was his habit.

Mark Dreher, Orbcomm's marketing chief, said a two-satellite business needed more industrial applications to break even. When Teleglobe signed in July, he explained, his staff was within days of signing its own agreement with Westinghouse to pursue five separate Orbcomm applications—monitoring refrigerated railcars, tracking trucks, monitoring remote assets, recovering stolen vehicles, and providing emergency road service. But without a firm commitment from the Canadians, the deal floundered.

"We'd been telling their people we had our financing in place, that Teleglobe's on board," Mark explained. "Then we find out Teleglobe won't put up $80 million and we have to cross all these hurdles first. We knew it would scare some folks away. All we could do at Westinghouse was tell them the truth."

The truth was that most of the services Orbcomm wanted to offer depended on a full constellation. With only two satellites it would be little more than a once-a-day data collection service—an entirely different business. Their prospects fell from a potential market worth a billion dollars to one that hit $10 million at best.

Doubts chilled negotiations with potential customers, and the marketing team scrambled to restore the company's credibility. Unless the full constellation started up before 1996, they believed, it would be hard to overcome the public perception that cellular services would be first to claim the mobile communications market, at least in the United States.

At one point Westinghouse had even assigned one of its executives to help Orbcomm win a product-development agreement with the American Automobile Association and Ford Motor Company. By forging a strategic alliance with Ford and AAA, it was thought, Westinghouse could offer emergency road service for all new Ford cars equipped with Orbcomm transceivers. Inside sources at Ford had been saying the automotive company showed definite signs that it would add new technologies like Orbcomm to its high-end luxury cars. In fact, plans were already in place at Ford to install a cellular communication system in the Lincoln Towncar by 1998. For the lower-end cars, the Taurus and Escorts, plans were less certain, but there was a possibility, especially with the Westinghouse connection, that Orbcomm could plug that niche by selling Ford an inexpensive, long-term solution. Naturally, a break into the automotive market could bring huge revenues.

Recently, in fact, discussions had been under way at Westinghouse to create an entirely separate division exclusively for Orbcomm product development. Some Westinghouse executives saw a natural fit between their services and satellite links to home security systems and refrigerated cargo on trucks, trains, and railroad cars.

But since the Teleglobe deal, Orbcomm's window of opportunity had shrunk. And that made everyone skittish.

"We've been handed a heavy set of handcuffs," Alan said.

The gentlemen from Teleglobe did not respond.

The good news, however, was that two small satellites developed by the Boulder team were now in orbit scanning their frequency bands, and the data showed that interference levels from the Earth appeared less intrusive in space than feared. This meant, of course, that Orbcomm's channel-scanning plan, created to identify clear RF channels for the Earth-bound Orbys to link into, had a better probability of succeeding.

Also, at that moment, downstairs in the Steeplechase parking lot, Orbcomm engineers were equipping a Ford Club Wagon with spectrum monitors and signal generators. Soon engineers from Orb Inc. would pair up for six weeks and drive one long circuitous route from coast to coast, testing noise in major cities, around industrial sites, and in the most rural, distant corners of the continental United States. The OTR ("on the road"), as David Schoen called it, would help them pass the first of Teleglobe's milestones. By early October they expected to prove that the little five-watt Orbys could withstand an assault of ground-level interference and block out RF noise generated by every possible interferer, from airports to hydroelectric dams to long-haul trucks.

The two test projects would cost the company a million dollars, Schoen said, but they would provide documented evidence to Teleglobe that Orbcomm could make a link from practically anywhere to Orbital's orbiting satellites.

The meeting was about to end on that bright note, when Bruce cleared his throat and suddenly changed the subject.

"I would like to purchase 200 Orbcomm messaging terminals," he said. Everybody knew immediately what he meant—search and rescue. "And I'd like you to place the order by next week."

"Okay, fine," David Schoen said, not looking a bit surprised. "But there's nowhere to place an order. You have to order them through Panasonic of North America, and they won't give them to you."

"I know we have a philosophical difference about these personal messaging devices," Bruce said. "But I want those terminals."

The blood in Schoen's face began to rise. Alan quickly jumped in.

"We don't have any philosophical differences," Alan said.

"Good. Then my question is, when do we get them?" Bruce said calmly.

"At the end of '94," Schoen said. "Just like we planned."

That was almost a year after the scheduled launch of the first two satellites, Bruce thought. With testing and production start-up, he would have to wait another year or two before seeing them on the market.

"I think that's just horrible," Bruce said bitterly.

"I'm sorry," Schoen said, "but Panasonic just will not sell a product out their door with their name on it until the satellites are up and tested."

"We need 200 terminals," Bruce said again, "and I want them before the first two satellites are launched."

Alan tapped the table with a finger and looked down at the floor. Ferguson had been pushing him for weeks about the messaging terminals—pocket-size devices with the QWERTY keyboard.

Problem was, Alan explained, the Orbcomm business was, at least for the moment, a "haircut version" of its former self, two satellites instead of thirty-four. With only two satellites Orbcomm couldn't deliver a message, in some cases, within six or eight hours. Besides, the company had no money to advertise a commercial messaging service. What they needed more of, and quickly, were industrial services, more customers for the machine-to-machine markets.

"Bruce, there's a certain logic to these things," Alan said calmly. "We'll never take our business from A to B without first clearly seeing what A is. And I can tell you, we can't start the business with personal messaging. You know, it's just very expensive to get the consumer market. You've really got to get your name out there to succeed. There's no magic to it. It's simply a matter of how much money you can spend for the acquisition of a client. And you might say, 'Oh, we'll do it for 5 cents per.' But that's bullshit. You're bullshitting yourself if you think that. You're going to lose. You're definitely going to lose. Later, perhaps, as the business grows and we get the full constellation going, then we can talk about Dick Tracy, but not now. This isn't the time."

Bruce shook his head wearily, impatient with excuses.

"Tell Panasonic to make us a demo," he said, looking Alan in the eye. "The sooner we do that, the sooner we get into production."

"Are you asking me now to go back to Panasonic and get a separate contract for a demonstration model?" Alan asked.

"Yeah," Bruce said.

"Then it'll cost you a million dollars!" Alan said, shouting across the table. "Because right now, Panasonic has fourteen of their people building these industrial terminals for us, and they consider that product development. Meaning, right now it's all on their nickel. We ask for a special deal, we pay full price. And let me tell you, that's a lot higher than you'd think."

"Besides," Schoen said, "those two hundred terminals wouldn't do anything to add to our revenue stream. It would be a distraction."

Their voices grew threatening. Alan threw up his hand to stop Schoen from speaking again and regained control.

"Bruce," Alan said, this time so calmly as to sound condescending, "Panasonic sees personal messaging as part of a full-up system. You can't do it with two satellites. It would have been a great investment, I agree, but we rocked our own boat with an inability to get the funding."

"I think you could say we showed an inability to perform," Bruce acknowledged. Alan's statement had sounded like a putdown, a veiled criticism about Bruce's inability to get Teleglobe to invest on the original terms.

"It's just reality," Alan said, shrugging. "The lack of funding has had a major impact on us."

"And don't forget who we're dealing with," Schoen said. "Panasonic has a very unusual relationship with us. We don't have a contract, per se. We have a good faith statement. You don't know how they do business over there. They will not build and sell anything until it's completely tested. They don't even have the ability to build two hundred terminals. They can't build *ten* terminals! Their scale is much larger than—"

"I insist that we apply the same ruthless standards to them that we do to any of our contractors," Bruce said, also quietly keeping his emotions in check. "I don't care about organizational problems."

Alan laughed. "That's great, but it won't get us the terminals."

"I also refuse to believe the only way we can get them is through Panasonic," Bruce said, his voice heating up. "Period. End of discussion."

Bruce scowled, folded his arms against his chest and, for a moment, looked invincible.

"Okay," Schoen said, "then tell us: What's your definition of a *terminal*?"

"I want something with a keyboard built in, and I want it before the first launch. Dammit! The public's perception of Orbcomm is a keyboard. It's not a machine-to-machine service."

"But that's our market!" Alan exploded, laughing because, at the moment, there really wasn't any *public*. "That's where our revenues are, Bruce! These are solid objectives! Industrial applications are the only real possibilities right now. And they're good enough that we had Westinghouse back here yesterday, and they were not looking at one single application that had anything to do with personal messaging. We can sell hundreds of thousands of units for machine-to-machine applications. And I don't see that as a trial business. That's a dollar-is-a-dollar business. On the other hand, there's no way to meet our financial requirements by selling all-out messaging terminals. I'm satisfied with that! With only two satellites you'd better believe however many of these little guys we sell for oil pipelines or potato bins will play a damn good role. You might find a few kooks out in the world who don't mind waiting six to nine hours to receive a personal message because they think it's neat to own their own satellite, but they're few and far between."

Alan crossed his arms.

"You're saying you can't do it?" Bruce asked calmly.

Schoen grunted. Alan looked like he would explode.

"It's one of the most important issues on the Orbcomm project," Bruce continued. "Can't we get someone else, another one of our other contractors, to build them?"

"Panasonic will build them, and they will deliver them," Alan said.

"On their schedule," Bruce observed with a sour look. "At the end of '94."

"Right, and that's a contractual issue," Alan argued. "It's already been decided, Bruce! The schedule's set."

"I'd like to accelerate their schedule then," Bruce said. "And we should be fair with them."

Again Alan laughed. "You better believe we'll be fair when we deal with them. There's no other way for us to be but fair! We can't be *unfair*. They'll just look at us and say, 'Sorry, we're not interested in you. Bye-bye.'"

Mocked, Bruce looked away. Then he leaned over the table and said, "Alan, my definition of *fair* is that we pay them more money."

He folded his hands and added: "I'm not going to walk."

Their exchange was becoming an embarrassment in front of the Teleglobe reps, who were there only to observe. Finally, Bruce and Alan agreed to meet later, privately, to again wrestle with the most basic and controversial issues in the business.

After two years it still was not certain exactly what the Orbcomm technology would do, what it would look like, and where it would be made.

Bulletin Board

"The problem is you need serious up-front investment—not an occasional test-bed of a few million dollars—to develop new technologies. And you have to have small missions to incorporate the new technologies."

—Lennard Fisk
Aerospace America

"Okay, who has the bad news? This wouldn't be a staff meeting without bad news."

—Mark Krebs
Satellite engineer
April 1993

"The issue is how do we carry Orbcomm and still preserve cash? The danger is that about the time we launch the satellites, we run out of cash. So we need to look at a ten percent staff reduction. At least go through the exercise. Make a list so we'll already have it in hand. The unfortunate reality is this first wave of cuts could hit sooner than we think."

—David Thompson
Executive staff meeting
June 1993

"We are putting almost all of our available cash into these satellites. We can't run out of cash. We don't want to run into a brick wall and we don't want to slow your team down. How will we do that? We're working on it. The company is also working hard to figure out how to organize the corporation so it will be more efficient for everyone. The organization we're a part of today was designed three and a half years ago under a different set of circumstances. So it's lived about as long as it should, and it's time to cook up something new."

—David Thompson
Conversation with satellite team
July 1993

GUIDANCE AND CONTROL: A spacecraft's guidance system, like the seaman who uses a sextant or radio navigation on a ship, keeps track of its own whereabouts. It locks onto its cosmic aim points and maintains its bearings relative to where they are. If a spacecraft sees itself wandering off course, it computers signal tiny thrusters to fire just enough to swing back again, the way sails on a ship are trimmed or the rudder is moved slightly.

SOLAR PANELS: "These panels pop open like flower petals and collect sunlight to power a spacecraft . They are connected to a "bus," which has tiny pitch, roll, and yaw thrusters to keep the satellite stabilized in all three axes, and keep its sensors properly oriented."

THE BRAIN: "The brain is a primitive computer and sequencer that sees that tasks are performed in order. A complex alarm clock triggers events after launch and converts the spacecraft from rocket passenger to independent spacecraft, opens solar panels, aligns it with the sun, and transmits data."

—William Burrows
Exploring Space

Top Ten Uses for *The Journal of Guidance and Control* on a Caribbean Vacation

10. Sun visor
9. Drink coaster
8. Makeshift battery suit
7. Anchor
6. Paddle
5. Plug leaks
4. Conversation piece at local bars: "Why, yes, I am a rocket scientist."
3. Wind direction indicator (wind turns pages)
2. Next best thing to a sleeping pill
1. Sobriety test

—Gift from Morgan Jones
to Mark Krebs
1993

Mark Krebs's Unified Theory of Human Behavior

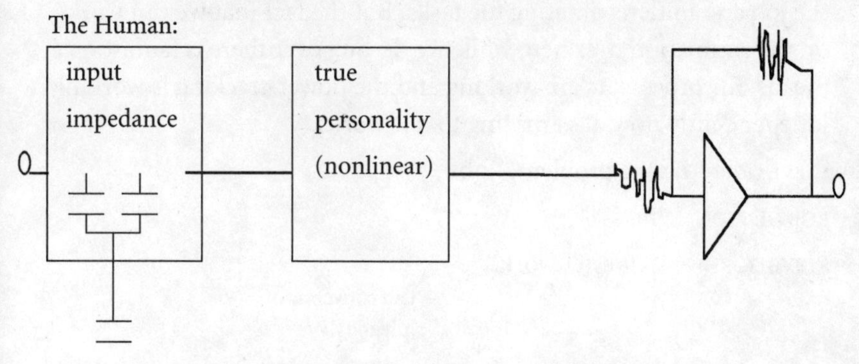

Model for a person has three elements: Input impedance, output gain, and true personality

CORROLARY #1: I/O characteristics typically define externally perceived type.

CHARACTER TYPE:

INPUT / OUTPUT	LOW	HIGH
LOW	Overly Sensitive	Tonto
HIGH	Prima Donna	Lone Ranger

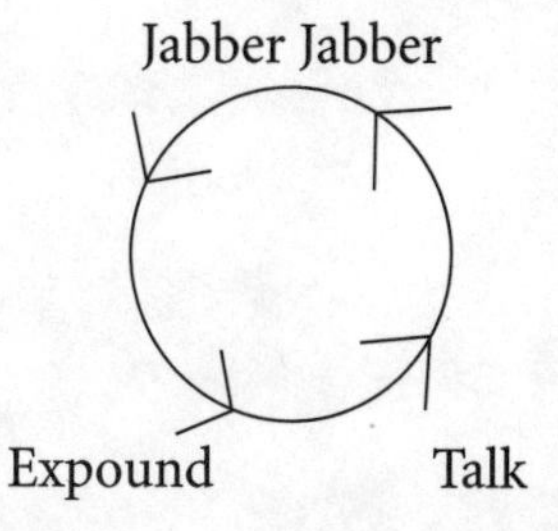

COROLLARY #2: To achieve inner peace (or interact with another's true personality), adjust input & output gains to neutralize everyone else's

EXAMPLE: The high gain/low impedance prima donna black hole, serial loop of PD's consumes infinite energy, never fully silent

—Mark Krebs
Gift to satellite team
June 1993

ROB DENTON: We have the classic problem of being undermanned for the job and underestimating the tasks. But the fact that we can work on the simulation over here while we de-bug over there is fantastic. It means our processors are working and the power-tracking is working independently now. It's amazing that it works.

STEVE GURNEY: One problem, Rob.

ROB DENTON: What?

STEVE GURNEY: It doesn't work.

—Lab conversation
July 1993

"The way to make a small fortune in space is to start with a big fortune. Space is a good place to lose a lot of money real fast."

—John Pike
American Federation of Scientists

CHAPTER THIRTEEN

August/September 1993

Yank 'Em

Late one August afternoon, at a bar near Sullyfield, the satellite builders joined Tony Robinson to "wet the baby's head," a ritual event in England, Tony's homeland, where friends gather to hoist a few beers at the arrival of a new baby.

At ninety-nine cents a bottle, the beer made several rounds and created a cheery mood as they toasted the satellite's chief layout designer, Sean Curtin, his wife, and their newborn.

Other engineers' spouses arrived, too, carrying babies in car seats, and while the spouses found lots to talk about, they eventually came around to common complaints: their husbands' many absences from home at night and on weekends; the lack of financial remuneration for dedicated work; an intolerable year of stress and sacrifice. They sounded unified, clearly not the first time they had voiced this litany.

The senior engineers sat ringed around two sets of tables comparing mortgage rates, and trying to guess who would bring along the next infant. They were boozy and a bit boisterous.

Mark Krebs claimed that under no circumstances would he and his wife, Suzanne, have a baby. "We don't even like the Orbital picnics," Krebs said. "More Blazers per square inch than anywhere else in northern Virginia. Disgusting! You can't even play a decent softball game for all the kids scooting around! Me with a baby? No way. I hate that purple dinosaur—what's his name?—Barney!"

Krebs and his wife had settled into a new home in Leesburg, Virginia, not far from Tony Robinson's house, at a 6.25 percent rate. The $50,000 bath on their house in California had hurt, but at least it had sold, and Suzanne was starting work as a schoolteacher. For the first time they felt rooted.

Eric Copeland was looking forward to his new Porsche, not the arrival of an infant. He also had some extra cash now that he and his wife had refinanced their house at 4 percent. Saluting the bearer of the lowest mortgage rate at the table, everyone clinked bottles to toast Eric.

They devoured a couple of trays of spicy buffalo wings, and someone ordered another round of beer. The crowd at the bar grew noisier as the dinner hour approached, and the engineers huddled together to be heard.

"I got a promotion," Tony announced.

Everyone looked his way.

"I was shocked, really," Tony continued. "Totally out of square."

"What's that mean financially?" Eric asked.

"Five percent," Tony said.

"Is that it?"

"And they gave me a new title: principal engineer."

"We should mutiny!" Eric said.

Looking around at the faces of the engineers, Tony could tell they were studying his news carefully. Probably not one of them had received more than a three percent raise at the end of the first half of '93. He knew Eric felt cheated—he had never earned less than 15 percent in his entire career at Hewlett-Packard—and just that day when he and Dan Rittman went to lunch, he had downed three beers to soothe those concerns.

"Suddenly, I've got the worst headache," Eric said, pinching his forehead. "But you know, I think it's time we did something. We've put up with this shit for too long."

Tony was silent. Krebs was working his jaw as if he were mulling over a deep thought. Dan Rittman shook his head and looked unhappy.

"Steffy doesn't stand up for us when he meets with upper management," Eric said.

"Come on, Eric, he's midlevel. What do you expect him to do but cover his ass?" Krebs said.

"I expect him to tell David Thompson we can't work this fast," Eric said. "We have to launch this sucker in February, and a lot of money is riding on us being ready. He should be finding a way to cut us slack, not yelling at us whenever we fall behind. We won't even come close to launching next summer unless every board makes it through testing. I mean, that's why we test! To find problems!"

Nobody on the team put in more hours than Eric. He had weathered one crisis after another with aerospace vendors who undermined his schedule by sending bad parts, producing sloppy boards, and missing deadlines.

Also, the stringent specifications set by Orbcomm for the spacecraft's communications gear had forced Jan King's group in Boulder against a wall, technically, and where they could no longer adapt to rigid specs, they called on Eric to help. By now, Eric was familiar with every corner of the satellite's communications system. A particularly tedious and uneventful game of redesigning receivers and transmitters highlighted the day-to-day nature of his struggles.

Still, he kept pace. Until testing began in July, he had managed so well as a troubleshooter that he earned even greater respect from the crew in Boulder. If Eric complained, he would be taken seriously.

"We've all been in and out of Steffy's office hounding him about these things," Krebs said. "I don't know what good a mutiny does."

"Then we should yank their balls!" Eric said.

An appreciative laugh went up. The men applauded, everyone but Tony.

"No, you can't yank them," Tony shouted above the laughter. "That's no good."

"Why not, Tony?" Krebs said. "You're always yanking Steffy's. You always give him a hard time."

"No, you've got him wrong," Tony said. "I think Steffy appreciates it when you're honest. He needs that to take to his bosses. That's the way it works."

"I don't see it," Krebs said. "I never see him stand up for us."

"He plays a good game," Eric said. "If he's doing something else, it's transparent to me."

"He always takes the company line," Krebs said, "even when you can tell he doesn't believe it himself."

"Exactly," Tony said. "He depends on us to set them straight. He appreciates it when we say something that he can't."

"What do you mean?" Eric asked. "Are you proposing something? I already said we should mutiny."

"No, you said, yank their balls," Tony replied, smiling. "I say just hold them in the palm of your hand very gently and then tell them in the most straightforward way, 'We want stock options.'"

The engineers let out with whoops and catcalls.

"I'll do it!" Eric said.

"Sure, we can ask for whatever we want!" Tony said. "A bonus—paid overtime—financial incentives—"

Tony and Eric pledged to start a campaign immediately.

"You've gotta be willing to lose," Krebs cautioned. "Don't forget what happened to Morgan. Morgan doesn't have a future with this company. He played hardball and lost—in the long run, he lost."

Everyone was familiar with the Morgan Jones story. During the summer Morgan had threatened to quit if the company refused to move him back to Boulder to be near his girlfriend. He had gambled by pressuring Orbital's executives, knowing they could not afford to lose his expertise on the troubled OSX operating system. When he finally pushed them into a corner and they moved him back to work in Boulder, it was clear that Morgan had accomplished the move, but he went without a future. When the layoffs came, he would be among the first to go.

"A couple of months ago," Tony said, "I'd have agreed with you. But Eric's right, now's the time. We could make something work for us now. We'd have to be careful. We can't just go in and demand something. We can't threaten them outright."

"No!" Krebs said. He slammed his fist down on the table and jumped up. "No way. Listen to us. We're sounding like a bunch of lame-ass steelworkers. Like a union hall of steelworkers. We're not common laborers, we're satellite engineers. I can't believe this! You guys can't be serious. We have to remember who we are."

His teammates looked surprised.

"You might not believe it, but we're all replaceable," Krebs continued.

He spread his arms across the table.

"Look, you're right, Eric, okay?" he said. "We can't keep working this hard. We can't. But everything else in this business is hard, too. There's nothing easy about what Orbcomm's trying to do. There's nothing easy about what the whole company is trying to do. There's not enough staff. No one has enough money. And time? It's ridiculous. But—"

At that moment the waitress slipped in with the check. Krebs sat down, red in the face, and upended the rest of his bottle.

Everyone was quiet for a moment.

"So what do you say we put this on the company's tab?" Tony said to the rest of them. "What's the damage?—Eighty-eight dollars. Not bad. We should do this every afternoon."

Eric was already standing, a grin on his face, his belly full of beer.

"I'm going to do it," he said.

"I'm with you!" Tony said.

"That's a three-beer threat," Krebs said.

Eric dropped a ten and some change on the table.

"Wanna bet?" he said.

He turned quickly on his heels, and soon the spouses and babies and other engineers disappeared out the barroom door after him. They left behind a sizable tip, and all went home to enjoy a few days off.

It was Labor Day weekend, after all.

Just as David Thompson had never intended to rely solely on space-qualified parts for Orbital's rockets and satellites, he had also never intended to create satellite and rocket teams with strictly space-qualified engineers.

Certainly Freshouts were fast and flexible, particularly inclined to ramp up quickly. On the Orbcomm team they epitomized the founders' goal of creating a company of Next Generation engineers with the best available talent. They joined the company at the top of their graduating classes, and because they entered as generalists, the cost of hiring them was relatively low. Although they had not been toughened or previously tested in the business of aerospace, no one doubted that they could withstand the pressures. Given Moore's law and the revolution in new materials, in programming, in commercial software, and in microprocessing, David Thompson could reasonably believe that if the first Pegasus rocket was flying inside his very own desktop computer at the time that, say, Mike Dobbles graduated from high school, then he could also expect in 1993 that any reasonably bright Freshout with determination and creativity could build a propulsion system or a power system or a transmitter or an antenna. In little more than a year, he imagined, a team of Freshouts could do just as the company's founders had—rely on whatever intellectual and creative gifts they had to compete in an industry that was generally thought to be choking on its own inertia.

But by the fall of 1993, it was clear that Orbital's executive staff had underestimated the technical challenge. As much as they might have believed in technological minimalism and the promise of a Microspace frontier, the young satellite builders had inherited a very difficult set of specifications.

Consequently, the satellite team began to grow quickly during the summer, as Orbital spent more money to hire experienced contractors and more veteran engineers into the project. The core team bucked up. Subtly chafing undercurrents of discontent were barely detectable—at least, during the working day, to a passing observer. But there was a slow shift in attitude that had nothing to do with the team's growth or the pace of their work or the natural rhythms of engineering that left them taking only one step forward for every two steps back. The change was not swift or bristly. It was not a change so much in attitude or perspective as in something less dramatic. The young wolf pups became noticeably less frisky. They rebelled quietly.

Up and down the once bright, unblemished corridors where Freshouts of the previous year had arrived idealistic and determined, engineers plastered absurd *Far Side* cartoons of cavemen rocket scientists and undernourished, pock-faced analysts in lab jackets. The salmon-colored carpets running the hallways grew dingy. A few fluorescent lights that had burned out were ignored, leaving the corridors dim and foreboding. Someone drew a large sketch on the whiteboard next to the secretary's desk of an Orbcomm satellite in pink, exchanging an elephant's curvy snout for the antenna and two large, limp ears for the solar arrays. The image remained there untouched, a sly insult whose sting was felt only as it lingered there week after week.

Posted outside Steffy's office, the continually revised testing schedule also attracted graffiti and a few mocking comments posed in unrecognizable script.

The dress code went by the board. Members of the team who once came to work in distinctive, fashionable clothes now wore khaki shorts and sloganeering T-shirts. A few engineers discarded shoes completely and scuffled around in stocking feet. Allowing such informality marked an effort to boost morale and encourage production. Black humor, even back talk, was tolerated.

There was hardly a sea change. But as modest evidence, small signs of desperation did represent a more regressive, less optimistic point of view and perhaps a new, more proletarian order.

It was no longer for the love of the company or for the good of the whole that Freshouts and their teammates spent weekends and evenings trying to harmonize long lines of code and improve the durability of their hardware. They did not necessarily consider the company's quarterly reports as they beat down ragged spikes simulating natural and man-made interference that broke through their electronic filters and left them sometimes wanting to beg for a more generous specification from the customer. They continued plowing into the work, Steffy would say, because they had hard problems: good engineers like hard problems. Sometimes, too, they still thought of themselves as accomplishing something important.

The little team of Freshouts and wolf pups with their new paradigms grew quickly but worked more slowly than David Thompson had ever anticipated. Their satellite posed many interesting dilemmas, after all, and threatened to beat them in a game at the edge of possibility every day.

To the untrained eye, an Orbcomm satellite looked simple. It weighed only 91 pounds and made a visually uninspiring first impression. Stretched out on a lab bench, it resembled a bass drum set sideways with two circular solar power arrays spread out like Mickey Mouse ears and a snoutlike antenna that protruded from the body. Unlike the large, complex, space-hardened electronic packages people expect to see inside spacecraft, Orbcomm's very simplicity, its diminutive size and weight, and its pedestrian design actually set strict parameters that made it an extraordinarily difficult machine to build. Essentially, it looked simple; but inside it was quite complex.

To begin with, the basic dimensions of the satellite were dictated strictly by business decisions. For technical reasons, the constellation required at least twenty-four satellites to provide blanket coverage of the Earth, with eight satellites in three orbit planes. Because launches entail a large part of the expense in space missions, it became financially imperative that the satellites be small enough to launch a full plane—eight at a time—all at once by stacking them like a roll of pennies inside a Pegasus fairing. Since the space inside the rocket's fairing was not much larger than the average telephone booth, all the electronic gear, harnesses, magnets, lenses, and fuel tanks had to squeeze inside a small cavity within each satellite body. The spacecraft also had to be light enough that a Pegasus could reasonably haul eight at a time into low-Earth orbit. The rocket could lift about 760 pounds into the desired orbit, meaning each satellite could not weigh more than about 95 pounds.

The financial decision to manufacture the spacecraft with relatively inexpensive parts also created technical problems. In most satellites engineers used expensive "radiation-hardened" or "mil-rated" parts so the satellite could withstand the Van Allen radiation belts above 1,000 kilometers and enjoy a reasonable lifetime. But a commercial project like Orbcomm simply could not afford to create a satellite network with mil-rated parts. That fact alone set the ceiling on the orbits at 1,000 kilometers. The satellites also could not orbit at less than 650 kilometers because residual traces of the Earth's atmosphere would create excessive aerodynamic drag on the satellites, make them difficult to control, and, ultimately cause them to reenter the atmosphere long before the end of an otherwise fruitful lifetime. So 650 was the floor.

Only by adapting to these technical constraints, set by financial considerations, could David Thompson afford to enter the global competition against much larger corporations. For a fragment of the cost of other constellations, he could produce a complex system for global communications and sell his services at discount rates well below the challengers. But the physical limitations and financial incentives also pressed his engineers to the extreme edge of their disciplines.

To begin with, the satellite was useful only if it could point the antenna accurately and consistently at the ground while it raced around the Earth. As a mobile system, Orbcomm couldn't expect subscribers to haul big antennas around with them on the ground, so it was critical that the spacecraft keep its antenna steadily directed at its target, no matter where it was in the sky. At the same time, to stay powered up, the satellite had to align its solar arrays toward the sun. Finally, as the spacecraft turned to hold attitude position for both the antenna and the solar arrays, it also had to angle itself in ways that prevented a shadowing effect by its own body from blocking sunlight on the solar cells. The total attitude requirements meant the satellite could not stray more than a few degrees. It all made for some very difficult gymnastics to move the satellite's body and appendages at various speeds and directions—a set of choreographic stunts that Mark Krebs worked endlessly to perfect.

On most large, sophisticated communications satellites, these kinds of complex contortions were made easy by apportioning 25 to 30 percent of the spacecraft's total weight (normally, hundreds or thousands of pounds) to the control system. On Orbcomm, however, such an allotment—about 25 pounds of its 95-pound total—was out of the question. The attitude system—the "cheap-ass system" that Krebs inherited—could weigh only a few pounds.

To keep the satellites in proper formation, without wandering out of orbit or bunching up in the orbit plane, the satellite needed a bottle of compressed nitrogen gas—Dobbles's "mouse farts"—to provide course corrections from time to time. In the first design the satellite included two such tanks, but because late specification changes caused other systems to grow in size, one of those tanks had to be eliminated. To make up for the loss and to provide more delicate adjustments, Dave Steffy thought up a way to "flair and feather" the solar arrays—that is, to turn them into or away from the sun—so that aerodynamic drag from sunlight could be used to move the spacecraft around as a complement to the gas tanks for position control.

As if these technical constraints weren't difficult enough, another complex set of demands challenged the communications systems. In the early 1990s, as the FCC considered ways to structure the new LEO frontier, Alan Parker and his advisers set the frequency band near 150 MHz—close enough to the upper range of FM radio to make it possible for commercial manufacturers to produce Orbys using cheap and plentiful mass-produced electronic parts. At the time, lower frequencies provided the best way to guarantee parts inexpensive enough for automotive markets.

The choice of frequencies, however, complicated the entire satellite system. The design and packaging of the satellite antenna, for one thing, became extremely difficult. If the frequency had been high-frequency L-band, a different kind of technology could have been used, such as is found today in cellphone systems. With a higher frequency the wavelength would have been short enough for the satellite antenna to fit comfortably and effectively inside the spacecraft and leave plenty of room for other components. But at VHF, where an efficient antenna has a characteristic length of several meters—much larger than the diameter of the Orbcomm spacecraft's body—the satellite engineers had to find a way to stow a long, foldable, extremely lightweight antenna into a narrow trough and then figure out how to spring it out once in orbit without any possibility of it snagging or breaking or hanging up. That was the responsibility of Tony Robinson—more of an art than a science, and one of the most difficult design challenges of all.

At last the communications aboard the satellite pushed the engineers right against the wall. The receiver, for example, had to be extraordinarily sensitive to communicate with Orbys on the ground, which were using small nondirectional whip antennas. It had to pick out Orbcomm's signals from a massive cacophony of background noises emanating from the Earth. Because there were no existing systems that matched Orbcomm's requirements at anything close to the desired weight, all the equipment had to be designed and built, with severely limited weight, power, and mass constraints, from scratch. That job fell to the team in Boulder and to Eric Copeland in northern Virginia.

In the end, all subsystems in the satellite posed a risk. Forced to maximize every available efficiency, the electronics, mechanical devices, and software grew more and more complex, more difficult to build, much harder to test, and nearly impossible to connect into one seamless, smoothly operating design.

From the beginning, business objectives and technical imperatives had clashed. A state-of-the-art satellite constellation—depending largely on passive designs and relatively inexpensive parts—looked fairly simple at the beginning of the project. But inside the little spacecraft's tight recesses, engineers fought to squeeze a highly evolved set of mechanical and electronic components that included some of the most sophisticated design technology ever seen in commercial practice.

One day in late September, the software team squeezed into a conference room to discuss their travail. They looked like a university class cramming for a test after hours. Annette Mirantes wore her Purdue sweatshirt, Morgan

Jones was in black shorts and black leather sandals. Steve Gurney pitched his Air Jordans up onto the desk and leaned back in a chair with a grimly defensive smirk on his face.

Somebody kept whistling the theme song from *Jeopardy!* whenever Morgan stepped up to the whiteboard to draw a diagram.

OSX was, by then, fully installed on each individual box, or subsystem, in the satellite. As usual, everyone demanded an alteration in the code. Although they had often worked communally at the beginning of the project, that had not been so true during the last few months, and the software design for each box, written by different people at different times over the year, suddenly developed a number of quirky operations peculiar to each engineer's private desires. Overall communication schemes aboard the satellite finally resembled a country with several warring dialects, and any effort to streamline it—to find a common language—made everyone testy.

Gregg Burgess, who supervised the full integration, listened as Gurney methodically critiqued OSX. Gregg leaned back in his chair and crunched on a carrot. Beneath his bloodshot and swollen eyes, the skin had turned purple from a sinus infection and lack of sleep. He looked sadly at Morgan, who stood tapping a red marker to his lips while Gurney finished a fairly devastating analysis.

For a moment a sense of doom hung over the room.

"Forget it," Gurney said suddenly. "Forget everything I said. Look, I've got a basketball out in my car."

They looked at him like he was crazy.

Just then the *Concorde* passed outside the window, on its way out of Dulles airport, and the engineers turned to watch it rise above the changing trees. Some of them had not, until that moment, even realized that fall had arrived.

"Great day for fluorescent light," Gurney cracked. As always, he simply had other things on his mind, and he wasn't about to let a beautiful day pass without getting his fair share of sunlight and fresh air. The software would have to spin off into the weeds without him.

Like many of his colleagues, Gurney had been born into an engineering family, grown up in the wealthy suburbs of D.C., and attended a high school where most of his friends graduated and left for good colleges. He had attended Washington University in St. Louis to study engineering, but many of his pals went to business schools and then jobs in New York. Though Gurney enjoyed engineering, he always thought he would follow his friends with an MBA and eventually settle in the City, too. Space was an interest, but not his ultimate passion.

Although he had the loping gate of a slacker and sometimes emoted a kind smart-ass attitude that particularly irked Dave Steffy, he really did have serious intentions for a career unrelated to hacking software. At some point, he thought, he would pursue an MBA and start his own company. For the time being, though, he was happy to work with smart people, take a ride with a back-of-the-envelope company like Orbital, and enjoy the good life of Georgetown.

With an apartment only a block from M Street, Gurney lived within a pleasant stroll of Georgetown's dining district and a particularly tony neighborhood whose hardware store offered help from salesmen who dressed in coats and ties and stocked a generous supply of hard-to-find metric bolts, which he needed for the endless repairs on his old Saab. His roommate, who was once his boss on a Nimbus satellite project at NASA, had moved out to live with his girlfriend, and Gurney had the entire apartment to himself for only $500 a month. He didn't care that the place was nearly falling down and threadbare or that his job sometimes drove him crazy. Georgetown was cool, life was good.

Since a number of his colleagues had gotten married or would soon, and conversations in the lab often swirled around mortgages and new homes or renovations of old property, Gurney decided sometime during the summer that he should begin to look at property, too. On a few weekends he cycled to run-down neighborhoods in northwest Washington, where he stepped through a few homes listed on the market. If he had a little more money, he thought, he could see renovating an old house.

Old things appealed to him: it was definitely the call of a smart retro style and philosophy, which assumed a kind of hip, nostalgic, inside track on the past—the glory, paranoia, bombast, clutter, adventure, excess, hype, and astonishment of American culture. The phone in his decrepit apartment was a black rotary dial, his car was used—the effluvium evoked by other eras smelled sweet to him. When Dan Rittman came to work one Monday morning and said he and his wife had spent the weekend remodeling their bathroom, Gurney discovered that they had destroyed the garish pastel pink-and-blue Formica and paint in their bathroom. He went ballistic. How could anyone demolish a classic example of '60s Mod? What were they thinking?

Before joining NASA as a contractor, he had also spent a good bit of time rebuilding old trains—not model trains, but actual zephyr locomotives that his roommate bought. Gurney would drive up to New Jersey on weekends and hang out for forty-eight hours in the train yards renovating old engines. His roommate eventually sold the antiques for a nice profit. Gurney never made much money with it, but the reward for him wasn't so much financial as simply the pleasure of sleeping in the yards and dipping into the guts of another era.

His work at NASA might have been a little like that—dipping into another era—but instead of having him tear down the old mainframes, one of his jobs was to install new equipment. He built ground stations for a while, worked on a Nimbus satellite, and wrote software for a couple of classified small satellite projects that, to his knowledge, never got off the ground. It wasn't just retro, at the time. NASA could be very boring.

"Just like working for the government," Gurney told me one day. "Things were unbelievably slow, and it was difficult to get anything done. I mean, I'm a software engineer, but I would go into work for NASA and sit there for months without a computer. It was like, 'Well, they're not in yet.' And I was like, 'Well, what do I do now?' I'd go hang out at the library."

Although the Nimbus project had some success, Gurney eventually strayed too far from the program management to enjoy it. "It was funny, you know. I was just sending and receiving messages at a ground control station, and most of the other guys they hired to run the controls didn't know shit about satellites or space or technology. It was just this job for them. Nimbus worked, then it died. It wasn't like I ever had some great feel for the satellite or got deep into the project."

Now Orbital couldn't have been more different. Here he could eat, sleep, breathe satellite if he wanted to. And nothing was very old. The company had spent $17,000 to buy him the latest Sun computer with a 19-inch monitor. Most of his teammates were about his age or younger. Everyone worked hard, and he had extraordinary freedom.

Best of all, the company had cachet. If the aerospace industry was something like Big Blue at IBM's overbloated zenith, where a lot of people managed managers who managed the hordes, then Orbital really was the industry equivalent of Apple Computer, reinventing products to be cheaper, faster, smaller, and more versatile. Elsewhere in the business, even among the best companies, engineers built spacecraft that would be three or four generations behind the last version of the most basic desktop computer. The design process, program reviews, testing, and office politics wasted time that a guy like Gurney simply could not afford to lose.

Then there was Cringley's law, which states that people who rely on computers for their work will not tolerate being behind the leading edge. Cringley's law applied especially to a guy like Gurney, providing another great advantage. At Orbital, engineers rarely suffered for a lack of the right tools.

Unfortunately, nothing was perfect, especially as the company continued to suffer growing pains. Sometimes Gurney thought the corporate culture had simply become silly and asinine, and sometimes he still had faith that if he stayed around long enough, Orbital would meet its larger goals—striking a gold mine, opening space to a new era of commerce.

If only he didn't have to conform to a homegrown operating system. That was the one issue that made life hard. If it had been up to him, they would be using commercial software.

After the meeting Gurney left the conference room with his Aspen coffee mug and stopped to climb a set of hard metal stairs up to the roof. He opened the steel plate on the ceiling and poked his head out, taking a refreshing gulp of cool air and lifting his face toward the sky. One of the technicians, who had spent the morning installing a set of GPS receivers on the roof, saw Gurney's head lift above the gravel and tarpaper.

"How's it out here?" Gurney called.

"Beauty-full," the technician replied.

Gurney looked around for a moment, ducked back in, and climbed down the stairs.

"Fluorescent light!" he grunted.

Shortly afterward, he went out to his car to retrieve the basketball and then disappeared.

One night I met Mike Dobbles and his Russian bride at their fifth-floor apartment in Vienna to take them out to dinner. Married for only a few months, the couple lived in an apartment that was sparse except for a few pieces of furniture, his books, and a pile of NASA technical reports that Dobbles had scoured from the Internet.

His wife, Angela, had spent the afternoon drinking coffee with a German friend and making calls about getting her real estate license. On the way to the restaurant, she complained about having to drive in his meager Ford Fiesta. She complained about his friends—aerospace engineers, naturally—who struck her as boring. He couldn't do anything about that. Finally, she barked at him about the way he was dressed—in typical Freshout attire, tennis shoes, Dockers, a T-shirt—while she, dressed fashionably in black, with her long dark hair combed to a brilliant sheen, looked as beautiful as a ballerina.

The tension mounted until we arrived at the Italian restaurant and settled in with a bottle of wine.

Dobbles was growing up quickly.

Through the year I had seen him complete the design for his thrusters, hire subcontractors, slice costs, burn off inefficiencies, and garner test results with the practiced nonchalance of a short-order cook. He had made himself invaluable by completing a piece of Seastar's satellite propulsion system, ingratiated himself with a manager in the Pegasus rocket program, and some days spent half his time thinking about designing rocket motors.

Orbital had been not so much his classroom as his campus, a place to improve practical skills and an intellectual environment that tested him daily.

At the same time, like almost everyone else on the team, he was also nearly oblivious to the larger scope and significance of how new technologies, and new satellite systems, had constantly intervened on his behalf. It was one of the real wonders of his situation, I thought, and yet whenever we talked, he seemed oblivious to it. All he seemed to know was that one day he was at home dying of boredom in Danville, Illinois, the next day he was studying in the Soviet Union, the next he designed a propulsion system for an experimental satellite, and the next day he was in Moscow meeting his girlfriend's family. He, the lowliest of lowly engineers: talking marriage with a Siberian girl; traveling with the company's top executives; arguing the usefulness of low-Earth orbits with the Soviet's master craftsmen in Moscow. First real adult romance, and he's jazzing it up with a vixen from Outer Mongolia. First job out of school, he's debating Ivan in a Soviet high bay.

What puzzled me was that he always acted as if his were the usual experience, a routine way of life. It was all so transparent for him. So much information and change appeared just within his grasp and control, so much of the future seemed to be just within his command.

His latest interest was finance. A lot of the books scattered around their flat, he said, had come from one of Orbital's MBA accountants, who loaned them out so Dobbles could learn the basics of accounting and business principles. He hoped they would help him understand how the company's financial wizards had managed to build an aerospace business without generating huge profits.

"I don't even understand a lot of those concepts—common stock and diluting outstanding issues and all that," he said. "It's like those first specification documents I looked at. It's magic. To me, the Orbital Sciences Corporation should be the Orbital Financing Corporation."

At last his wife smiled. Angela was completing her college thesis for a professor in Russia, using Orbital as an example of a successful capitalist venture. The thought reminded her of their romance.

They both remembered being in the company of Bruce Ferguson, the company's chief operating officer, when he brought Dobbles along as an interpreter on their trip to Russia the previous year. The night Angela met them at their hotel, as they were walking through the lobby, Bruce was addressed by two Russian prostitutes.

"You want sex?" one asked temptingly. "It is good for you."

Bruce replied: "No, no thank you, but thanks for thinking of me, anyway."

They laughed and told stories while we ate, recalling the halcyon days when they met on a college exchange program. They talked about their long-distance romance and their phone conversations during the collapse of the Soviet Union, when Angela and her friends stayed awake for three straight days in Moscow watching CNN, intoxicated by the sight of the Soviet White House burning.

"Isn't it great, Mike!" she would say, as he heard the news on television in Virginia. Linked, interestingly enough, by satellites, as the Soviet Union crumbled, they anticipated the day when she would come to live with him in the United States.

During the year while he was building a satellite propulsion system, he also schemed to bring the Siberian girl to America to be his wife.

"I like that he is so confident," she said, finally looking at him affectionately across the table.

That much was certainly true. He had grown in confidence during the past year. After completing the propulsion system, his skills as a problem solver became quickly apparent, and other program managers called on him to work in other programs.

As a result of his success, Dobbles had undergone something like a full religious conversion to the company. His interest in the business grew at a rate comparable to his love of engineering. He conducted research on the company, ordered several Harvard case studies about Orbital to review, sought financial prognostications from industry analysts, and took an interest in the company's dealings with Wall Street. He even began to spend a little time with some of Orbital's own accountants to see if he could begin to get the inside story of just how David Thompson and Bruce Ferguson had built an aerospace company out of thin air.

Every aspect of Orbital's business plan seemed brilliant to him. In fact, when the master plan became clear to him, he imagined himself creating a space business someday.

Angela seemed to enjoy listening to him talk about creating his own company. His most recent plan, concocted with one of the lead Pegasus engineers, was to build a commercial launch vehicle that would rival the Pegasus and sell for half the price.

Unfortunately, the day when he might leave looked nearer than anyone might have ever imagined. Over the past few months, some aspects of the business troubled him. Unlike Gurney, who never tended to worry about anything in particular, Dobbles could fret in a big way, and he could be especially critical when things didn't work out just as he expected.

During the first half of the year, for example, when Thompson announced that Orbital would no longer pay for professional dues, stopped the free supply of soft drinks, and asked employees to take on a greater share of insurance coverage, Dobbles joined his colleagues in the usual bickering.

I didn't think much about it until he called me one day to his cubicle and told me what he had just done.

"I sent an e-mail to one of those bean-counters at Steeplechase," he said. "Some guy—the business manager. I never even knew we had people like that until they started nickel-and-diming us."

He tossed the words around casually, with a distasteful spin—*bean-counter, nickel-and-diming.* I had assumed, at first, that he had done what some of his teammates did and simply fired off a note to Thompson or complained to Steffy, then let it go. But it became clear that he had thought a lot about the news and concluded that something was terribly wrong. He took seriously Orbital's open-door policy, and decided to call on the head "bean-counter," the company's vice president of business operations, Jim Utter.

To his credit, Utter received the young Freshout politely and offered to explain more about the company's financial predicaments. If he could not allay everyone's fears, Utter told him, he could at least try to provide Dobbles with a little background on the decisions that cut insurance benefits and halted the flow of free sodas and money for professional memberships.

"I used enough restraint not to call him a bean-counter to his face," Dobbles said, "but it was obvious he didn't know anything about engineering. I don't even think he knew where I worked. In fact, I bet you he couldn't find my office right now if he had to."

Clearly, Dobbles was wounded.

"I told him that Orbital's got the best employees, which is true. I told him that David Thompson and Bruce Ferguson built this company with a vision—that entreprenuership in space can work. It's this vision and also because David and Bruce don't expect us to be like everyone else—that's why we give 110 percent every day. He looked at me like he didn't know what I was talking about."

Apparently, the "bean-counter" told Dobbles that it cost the company something like $15,000 a year to pay professional dues and maybe the same to keep refrigerators stocked with sodas. Dobbles could see that he was intimately familiar with the budget's line items, but he could not force Utter to make the connection between that piddling chump change for Cokes and the ideals that motivated engineers like him "on the working level."

That, Dobbles said, was "really interesting."

It was a wonder that Utter hadn't simply sent the Freshout back to his cubicle or told him to take up his complaints with David Thompson. But Utter hung in, and in fact he finally chose to be even more candid. Simply put, Utter told him, Orbital's executives were scared. "They don't know where else to turn," he said.

Then he revealed something that only the program managers and executives knew—if the company didn't find more business soon, the rank-and-file faced layoffs. Rather than threaten or provoke the workforce with that kind of news, Utter said, management chose to cut insurance and quit paying for eyeglasses, professional memberships, and sodas.

Then he told Dobbles something that really upset him.

"He said, 'They're out in the middle of the ocean, and they don't know which way to swim!'"

Dobbles paused, and I could see he was waiting for my reaction, as if I would find Jim Utter's response as absurd as he had.

The comments only reflected the confusion I had witnessed in executive staff meetings since springtime. I didn't feel like it was my place to comment, so I shrugged.

"Well, we're not going to back down," Dobbles said defiantly. "They can threaten all they want. Everyone on this team is putting out a major effort."

In their meeting Dobbles pointed to the apparent contradictions—the executive staff had won a hefty pay increase from the board of directors that year, they had built a beautiful new headquarters and high bay at Steeplechase, they had won the support of Teleglobe.

"How can you turn around and act like we're in financial trouble?" he had asked.

"'I'm only one person,'" Utter said—another comment that still rang in the Freshout's ears.

"Can you believe that?" he asked me, this time not waiting for an answer. "And I'm thinking, 'Yeah, but you're a corporate vice president!'"

So he had just decided to take his complaint one step further up the chain.

"Bruce Ferguson and David Thompson don't want this," he told me. "I'm not convinced they really know what's going on."

The idea that Orbital was still a kind of family played in the young engineer's mind. He had typed out a personal letter to Bruce Ferguson and outlined his concerns. "Oh, I know Bruce will respond," he had said. "We can talk. He's one guy I have a lot of faith in."

Several days passed, and then weeks, and Dobbles still had no response.

He no longer wondered why he had never gotten a reply.

At dinner Dobbles and his wife talked often of their dreams, of engineering and rockets, corporate finance and new ventures in space. The niggling problems at work and at home seemed to pass from their minds for a while, and when we left the restaurant and walked out to the car, they were holding hands.

They both seemed happy by the time I returned them to their apartment, as if everything would work out after all. For them, if not for Orbital, anything was possible.

The whole team had become intimately acquainted with the new unwritten laws of commercial aerospace. Vendors and parts suppliers who operated under the old rules of an industry that had, for decades, grown accustomed to fat federal budgets and long lead times, did not always place a premium on timeliness in filling orders or delivering goods at the lowest price. Pressed by difficult specifications, the Freshouts made mistakes designing their systems. They forgot to fine-tune every vendor's contract, and learned the hard way not to assume anything that wasn't spelled out in legal documents. No matter how swiftly and earnestly they sprang back from their errors, the fact was, they simply could not launch until the middle of 1994. Maybe later.

One day at the senior staff meeting, Dave Steffy reported on a meeting he had had with Orbital's executive staff. Problems with morale on his team and others remained widespread enough that David Thompson had set up a plan to address it. When findings of the company's first report on morale was read, Thompson made the comment: "It's great to act like you're Ben and Jerry's if you're in a business that has 40 percent profit margins. And if we want to sell premium ice cream, then that's the way to do it. Unfortunately, that doesn't happen in aerospace very often."

Steffy recounted Thompson's words, and suddenly Tony Robinson leaned across the table to looked his boss in the eye.

"Has David Thompson changed as a person since you've known him?" Tony asked.

Steffy paused a second. He must have known Thompson long enough to have an opinion, though he was wise enough never to make that opinion known to anyone on the team.

"Well," Steffy said, "I'll tell you what he told us. He said he's really concerned that this is not the company he set out to create."

One of the engineers laughed, another murmured something inaudible.

More often than not lately, Steffy reached moments like this, trapped between his engineers and his boss. He trod carefully down the middle, often drawing back to cite a lesson in finance.

"The truth is, people are sincere in every direction," he told them. "But you try to find some way to improve the situation that won't run smack into a profit constraint. It's hard. Changes in the medical plan will save us $900,000 this year. That's pure, white-as-snow profit. And if you look at last year's numbers, you'll get a feel for how significant that is—that's a third of last year's profit. We have to recognize that we're working in a marketplace and in a stock market. Maybe we could manage better. Maybe that would raise the stock price. And maybe David Thompson should read his own e-mail so he'd know how everybody feels about—"

But he caught himself and paused again.

"I don't know," Steffy sighed. "There's a theory that says managers have all the control. That's bull. I can give you a hundred examples a week where I don't have control of this program."

"Then we should give you *more* control," Tony said.

"Maybe that's right," Steffy said. "I've tried to give you guys—"

"No, you give plenty control to us," Tony said. "I'm talking about you! You're the manager. You need more control."

"Except my job is to take your heat about the company," Steffy said. "That's part of my job description. And another part is, you've got to hear from me about what the brass wants."

"Well," Tony said, "that's frustrating."

"Yeah, it is frustrating," Steffy said.

And with that, engineers around the table began to rage and shout.

Not one of them was yet willing to believe what seemed abundantly clear to the executives and to the bean-counters: At the moment, and for some time to come, the fate of the Orbital Sciences Corporation rested squarely, almost entirely, on the shoulders of one small team of young, overworked engineers.

Bulletin Board

"Scholars argue that the development of a new technology involves a process of social construction by which a community of individuals and organizations negotiate a definition of the goals and character of the new technology."

—Pamela Mack
Viewing the Earth

"The thing Denton and I were really bummed about was, the upper management has been giving themselves these huge fat bonuses, 20 to 25 percent, and people at my level and below got exceptional reviews but our bonuses were flat. At the same time they're slashing benefits and taking away some of the things that made Orbital a good place to work. That's what we went after Dave Steffy for. It was a matter of principle.

His response? A resounding silence. I haven't heard a damn thing. Nothing."

—Eric Copeland
Satellite engineer
July 1993

"You hear a lot of these managers around here saying, 'I can't wait until these satellites get up. We'll all be rich.' I'm thinking, 'we' who?"

—Steve Gurney
Satellite engineer
July 1993

"The major problem for me right now is how to keep all these young employees fired up. They've made a career commitment and there's a lot of worry about their jobs, and we can't seem to get more capacity for the future. What do you do about that? When there's a survival mentality, the way the market's going, how do you respond to that? It's the worst job I've got, planning for the future of these people. How do you go to someone and say, 'We've got to let you go'? I've been struggling with it. It's just not a pretty picture right now. This is just not a fun place to work anymore."

—Bob Lovell
Space Systems president
July 1993

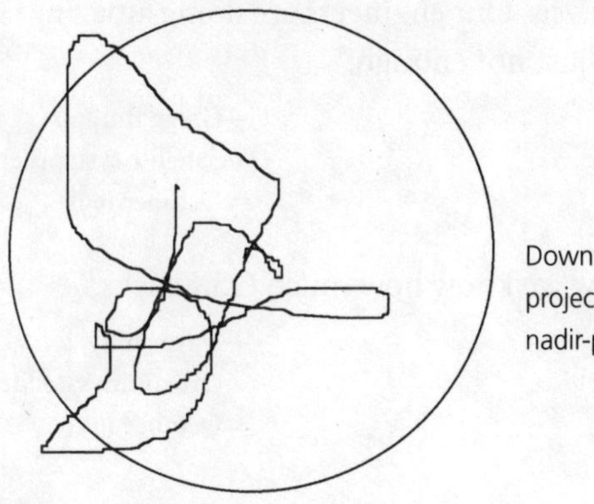

"This is your brain on Orbcomm. Any questions?"

—Sign posted in Mark Krebs's office
September 1993

"This hasn't been getting as much press coverage as the big LEOs, but it is vitally important in those areas of the world where there is no communication infrastructure. With respect to the big LEOs, $3 a minute for telephone calls will not be affordable to people in the outback or in a remote village in Africa."

—Alan Parker
Space News
October 31, 1993

"Don't give me shit, there's blood on every one of those fly wires, man."

—Rob Denton
Satellite engineer
October 1993

"What we have done with this spacecraft is we have avoided doing a very robust upfront analysis by relying on a robust configuration. But the only way to do a fly-on-the-edge system like this is to do the upfront analysis. Our engineers are doing amazing stuff on these satellites but it's just not enough."

—Gregg Burgess
Satellite systems engineer
October 1993

"At least now we know how much OSX sucks."

—Dave Steffy
Orbcomm satellite program manager
October 1993

CHAPTER FOURTEEN

March 1994

The Gentlemen from Caracas

The two gentlemen from Caracas arrived for the first time in March, driving a rental car, wearing light spring jackets, sliding across bumpy patches of ice in ten-degree weather.

Coming to a halt, a few feet from the front door of a small cinderblock house, they climbed out and stomped through a hard pack of snow, circumnavigating the gray blockhouse where a small engineering crew was holed up. The gentlemen whispered as they turned in the blinding sunlight, clouds of vapor rising out of their mouths and nostrils, glancing in every direction at expansive blue vistas and white landscapes from the mountaintop above the town of Arcade.

"Estamos aquí," one of the men said at last.

Passing in front of the white radome, the taller one handed a camera to the shorter one, who paused to frame a portrait of his friend against the white panels. The tall one said it was his first time in snow and grinned boyishly.

Inside, the Earth station's horn antenna turned like the minute hand of a clock, slowly locking onto satellites whizzing hundreds of miles overhead. A brisk wind swept across the surrounding field, vibrating dry, golden stubs of cornstalks, rattling a ventilation screen in the bunker's window. The men finished taking pictures, hurried over to the cinderblock house, and stepped in out of the cold.

Stan Ballas met the Venezuelans at the door but didn't bother to shake their hands. He had spent the morning at the hospital, where a doctor stitched up a gash in his palm, suffered when a metal shelf fell on him as he tidied up the bunker for the first series of antenna tests.

"We inaugurated the Earth station this morning," Stan said, holding up his bandaged hand. "With *blood*, sweat, and tears, you know."

"Welcome to Arcade," said one of his engineers, Dave Figueroa, standing up from his computer station. "Glad to have guests—it gets a little lonely up here."

Outside, a dusting of snow swirled against the house. A Federal Express truck rumbled up the icy path along the mountainside. Inside, the air felt perfectly climatized, a comfortable balance between temperature and humidity to protect the electronic control boards and computers tracking satellites with the cone antenna next door.

The Venezuelans unzipped their jackets and looked around the crowded room.

Their attention turned to Figueroa 's laptop, booted up into an amber screen, which displayed a map of the world. A footprint of the small satellite launched by the team in Boulder, OXP-1, rose as slowly as a soap bubble north from Mexico in their direction—elevation minus 12.2 degrees, altitude 769 kilometers. Next to his mousepad Figueroa scanned a list of other satellites they had tracked through the morning—weather satellites NOAAs 9, 12, 13, and the other Orbcomm experimental satellite built by Jan King and his Boulder crew, OXP-2.

Having served their purpose by scanning for interference in the Orbcomm bands, the little OXP satellites now gave Alan Parker's engineers a way to test out the company's first Earth station. For days they labored intensely, revising software that commanded the antenna pedestal to turn, measuring interference levels along the horizon, knocking out the power in the station to see how well it recovered, and testing RF links with modems that represented Orbcomm satellites. Every once in a while, they would call the network control center back in northern Virginia to report their findings.

Figueroa reached for the latest version of Orb Inc.'s VXWorks chip, a piece as small as a thumbnail, and popped it into a Gateway Earth station board marked for that very day, 3/11/94.

"Hey, somebody go hook up those GPS antennas on the roof," he said.

The only other engineer present, Figueroa 's boss, Steve Mazur, looked up from his computer, rubbed the stubble on his face, brushed the hair out of his eyes, and glanced at icicles hanging from eaves outside the window. They both looked like grungy Arctic fishermen at the end of a long week.

"We've only got four minutes before the pass," Mazur said. "GPS can wait."

The gentlemen from Caracas squeezed in to watch as OXP-1 rose slowly on the world map. The rim of its 3,000-mile, egg-shaped bubble edged closer to the bright orange dot representing their little blockhouse in Arcade.

For the next few minutes they stared at the animated screen, hands in pockets, shifting from foot to foot, rarely speaking, never looking each other in the eye, waiting for the antenna next door to lock onto the pass.

"There it is," Figueroa said. "Eight degrees elevation."

"I better call David Schoen," Mazur said.

Santiago Aguerrevere and Ariman Sanchez laughed and spoke excitedly in Spanish. The Orbcomm project actually did exist somewhere other than on paper. The Earth station's tracking system—at least on this end—came blinking to life. After waiting four years, they thought they finally would have something real, something palpable, something *spacelike*, to tell their doubtful backers at home.

But as OXP-1 rose 9.7 degrees over the horizon, the large horn antenna next door turned to hold its signal, and the abrupt motion left it wobbling unsteadily behind the cover of its great white dome.

Momentarily, the little satellite lost contact.

After months of relentless investigation, problems with unmet specifications, and slipping launch dates, Teleglobe finally invested another $8 million in the project.

For Santiago Aguerrevere and other potential investors like him, the Canadians' continued investments boosted their faith. Aguerrevere had waited four years and could not afford to stall his own investors at home much longer.

He had met Alan Parker through a mutual friend in Washington in 1990, an encounter that seemed, at the time, serendipitous. Venezuela's government-owned telecommunications industry had privatized only the year before, and as president of Telecomunicaciones de Caracas, Aguerrevere anticipated a significant market in his country for an inexpensive mobile satellite system like Orbcomm.

Venezuelans had hungered for a proper wireless service for many years. Any reliable communications device that could overcome the dense jungles and mountains where landlines never reached would do a handsome business in his country, and the prospect of using satellites, he predicted, would be greeted well.

Impressed by Parker and the new company, in 1992 Aguerrevere demonstrated his support for Orbcomm by traveling to Spain to attend the World

Administrative Radio Conference, where he lobbied effectively on behalf of the business. He was in a good position to argue that a new slice of spectrum, set aside for an inexpensive system like Orbcomm, would be a boon to the world's developing nations.

At the time, Latin America appeared as a tantalizing target in Orbcomm's marketing plan because of the overwhelming need to modernize the region's communications infrastructure. U.S. telecommunications exports to Venezuela had already leaped 80 percent from 1990 to 1992. All across Latin America, as more countries shifted from government-owned phone and electric systems to private services, Latin American companies and U.S. firms built alliances that would make Latin America a prime market for American equipment and services. In 1994, just as Teleglobe agreed to continue its investment in Orbcomm, Latin American nations, on the whole, were prepared to plow $12 billion into new technologies that would upgrade outdated and floundering telecommunication systems in their region.

After delegates to the world conference agreed to provide spectrum, Aguerrevere signed on enthusiastically as Orbcomm's first "candidate licensee."

By early 1993 there were nineteen more countries across the globe where private companies expected to build Earth stations and develop systems to take advantage of Orbcomm's new technology. They had been told that a full constellation of Orbcomm satellites would thread bits through the heavens by 1995, at least three to five years before any other LEO system in the world.

In Venezuela, Aguerrevere told leaders of government-owned oil companies that he had found a solution to some of their worst problems—a way to communicate easily and cheaply from the jungles to the sea. He convinced the country's trucking companies that they would soon be able to communicate with their drivers when trucks broke down in the jungle or on remote mountain ranges. Given what he knew at the time, he told them he fully expected to bring a revolutionary telecommunications business to Venezuela in just three years.

Of course, by 1994 the launch of Orbcomm's first two satellites, FM-1 and FM-2, had drifted two years beyond schedule. Even if the first two satellites orbited in 1995, it was clear to Aguerrevere that a two-satellite system would not meet a nation's needs. Two satellites would provide only ten-minute passes eight times a day in Venezuela, hardly adequate to serve the country or turn a profit for Telecomunicaciones de Caracas. The tentative nature of Teleglobe's investments, which became increasingly clear, also created doubt.

To make matters worse, the delays came with a surprising increase in costs. To those who were not privy to executive staff meetings inside David

Thompson's office, the reasons for the sudden hike in prices did not always make sense.

As the year began, staggering losses in defense dollars for aerospace companies had undermined the financial security of Orbital's Arizona offices, where engineers had been designing and building the first gateway antennas for Orbcomm Earth stations. As a wave of layoffs rocked Orbital's southwestern offices, the word came from Arizona that building an adequate gateway antenna proved much harder than expected and the need to hire more consultants increased the development costs. Although some executives in northern Virginia—particularly Alan Parker—believed Arizona's executives passed along the cost increases as a way to build its coffers in the wake of layoffs, there was no way to stop the run-up. By 1994 Parker saw that he could no longer afford merely to lease an Earth station to their friends in Venezuela, as originally planned.

So the deal changed: Santiago Aguerrevere was told he would have to buy—not lease—an entire Earth station outright at a price of $3 million, nearly twice its estimated cost in 1992. Consequently, given delays and price increases, what had once been a good relationship turned speculative and tenuous.

Just days before Aguerrevere and his colleague flew to New York to observe the gateway tests, I spoke to Mario Florian, Orbcomm's Latin America sales manager. He was bristling with anger.

"We are not being fair to Santiago or his company," he complained. "After three years of tremendous support from him and his people, we insult him by telling him he will have to pay for *our* development costs! He's also facing a new management here—Orbcomm is no longer monolithic. He can no longer deal simply with me. If he wants to talk, he has to go to Alan, Bruce, David Thompson, *and* the Canadians. If we delay much longer, I am afraid he will see that there are other options waiting for him, just like all of our other friends in Latin America. I'm afraid we're going to lose his business."

Repercussions from Dave Steffy's flagging launch date extended far beyond its effect on Orbcomm sales to Venezuela, though. The pressure of competition had begun to squeeze Orbcomm from several directions.

In 1994, for instance, the cellular phone industry in the United States began the year signing 14,000 people a day for wireless services. Its growth rate had exceeded 30 percent a year. Once driven by businesses wanting to improve the productivity of on-the-road employees, cellular phone companies had seen an enormously fruitful shift to consumer sales. Cellular services had swollen into a $11 billion-a-year business and showed no signs of slowing down. The average monthly bill had fallen nine percent a year since 1989, and plans to expand coverage beyond metropolitan areas into more rural and out-of-the-way places—always a potential threat—sounded more serious than ever. For Orbcomm, as with all other proposed mobile satellite systems, cellular radio promised keen competition.

It was no great surprise that Orbcomm's head of marketing, Mark Dreher, left the company at the beginning of the year and went back West to join the cellular business.

As the satellites' launch date slipped, too, Orbcomm's lead over larger LEO systems, such as Globalstar and Iridium, began to collapse. One of Orbcomm's most significant advantages, Alan Parker had often said, was its three-to-five-year headstart over the big LEO systems. Whoever broke into the skies first had the first-to-market advantage, allowing that system to set the standard for communication protocols that every other follow-on company would feel compelled to use. If pressed, Parker would shrug off the big LEOs, saying they had set their sights on entirely different markets—Iridium and Globalstar, he would say, had targeted a "high-end" user, globe-trotting business executives and their ilk, who needed voice communication, not simple messaging. But that was never a certainty, and in fact over the long run, industry pundits predicted the future of the business might favor an Iridium-like system, since it would offer voice as well as data communication.

Most of the time Alan Parker tenaciously avoided the topic. But others were more blunt and outspoken. Even the satellite engineers sometimes predicted that "Orbcomm will either be a smashing success or go the way of the eight-track tape"—an analogy that by 1994 no longer struck anyone in the company as particularly funny.

By 1994, too, big LEO systems surprised the financial world by attracting enormous sums of money. That month, for instance, Globalstar announced it had raised $275 million in equity financing for its $1.8 billion system. With substantial support from Motorola and Lockheed, Iridium executives boasted that they had already raised $800 million of their $3.4 billion goal and they anticipated another $700 million in new equity capital from corporate investors in Japan, a group that included Sony, Mitsubishi, and Mitsui. Iridium's other supporters included: Sprint, with an investment of $70 million; Veba, the German electric utility and phone service, which invested $140 million; another $32 million was expected from a consortium of financial institutions in India. (An important source of funding, India, with 886 million citizens and a burgeoning middle class, was considered a boom market for wireless communications—the consortium's investment spoke reassuringly to Iridium's executives about their opportunity to seize a lucrative share of the market.)

The enormous financial gaps between Orbcomm's costs and its competitors' no longer seemed as significant as they had only two years before.

There were other pressures, too, created by the sudden influx of inexpensive Soviet missiles onto the world market. In the economic chaos at the end of the cold war, former Soviet companies began to offer select missiles from their stockpiles of ICBMs at cut-rate prices. The temptation for Iridium executives to convert former Soviet weaponry into rockets that would launch their seventy-seven-satellite system appeared too sweet to resist. Despite angry protestations in Congress by David Thompson and other executives at Orbital that the competition exposed privately funded rocket businesses in the United States to devastatingly unfair competition underwritten by foreign governments, Iridium executives went ahead with their plans. For their launch services, they selected Krunichev Enterprises of Moscow and China Great Wall of Beijing.

The men and women in Alan Parker's marketing department were shaken by the continual launch slips. Not only had they seen the sudden departure of marketing director Mark Dreher, but they also had to keep explaining to clients why the constellation would not be a full service by 1995, as promised. At the same time, with no assurance from the satellite team, the marketing crew puzzled over how best to respond to questions from customers who were growing increasingly edgy.

But the most intractable and most threatening issue of all resulted from events occurring on a much larger scale: the entire aerospace industry was in free fall, putting the Orbital Sciences Corporation at serious risk of disintegration.

With defense procurements slashed in Washington by 67 percent between 1987 and 1994, even the world's largest and most successful aerospace companies had begun to falter. At Martin Marietta, the popular and widely respected chief executive, Norm Augustine, sounded an alarm: "These are Darwinian times in aerospace," he told members of the space industry press. "Our market isn't just declining, it's truly collapsing."

Aerospace executives, on the whole, estimated that the industry was as much as two-thirds larger than it needed to be. Massive layoffs regularly sloughed off well-educated engineers into lesser-paying jobs. (Home Depot, for instance, had started hiring out-of-work aerospace technicians on Long Island to help customers with home improvement projects.) As competition for smaller and fewer defense-related contracts grew intense, talk of mergers and acquisitions, even among the most prominent aerospace giants, stoked a mania for powerful synergisms that would not merely shore up hobbled corporations but link them and shape them into newer, stronger, more commanding, more attractive versions of their older selves, offering efficiency and vision—*interoperability* was the code word—for the twenty-first century.

In fact, corporations of all kinds throughout the United States entertained the same notion and went on a search for that special combination of expertise in defense, telecommunications, financial services, transportation, food, and entertainment that would turn megacorporations into megamonolithic entities of monumental proportions.

For example, in 1994 the value of announced mergers and acquisitions in the United States would climb to a record $111 billion, outstripping the old mark set in 1983. Top managers in the most prominent corporations were said to have compiled lists of the five companies they most wanted to acquire, expecting that a timely, strategic merger promised a quantum leap ahead of the competition, made the best use of available cash, and improved their chances for survival in a fast-changing marketplace.

In aerospace mergers and acquisitions made especially good sense. Defense companies traditionally operated in what was called a "cost-plus" pricing system. That is, on any particular proposal, a company would project all its costs, tack on a profit, and present the bid to Uncle Sam. If six contractors bid the same program, but each was already saddled with a number of underused or idled plants, then the bids would reflect the higher unit cost that resulted from their company's own inertia. But if companies merged, as some would, aerospace executives discovered, they could close more of those idled plants, cut their workforce, lessen the competition, and take greater risks by bidding lower. Also, since most companies spent a healthy percentage of their income on research and development, those duplicated efforts could be combined, at yet another savings. It was no wonder that the first of Norm Augustine's aphorisms was: "The best way to make a silk purse from a sow's ear is to start with a silk sow." It should have surprised no one to learn that by spring 1994 officials at Augustine's Martin Marietta company, the nation's fourth largest defense contractor, and at Lockheed, the nation's ninth largest, planned a $10 billion merger, the most remarkable confluence of defense-related companies in history. Augustine had set out, in a sense, to create his own gargantuan silk sow.

Orbital Sciences, of course, was not immune to the same pressures or unwilling to entertain the same fantasies. In 1993 it posted twenty-fifth in total income among the world's most prominent aerospace companies. The business was still small enough to be vulnerable, particularly as news and rumors of mergers like Lockheed/Martin seemed likely to change the competitive environment entirely. To avoid a possible takeover or, more likely, to escape being crushed by merging giants within their industry, Orbital's executives began planning the company's future development with a shrewd strategy of mergers and acquisitions.

In September 1993 Orbital had acquired a division of Perkin-Elmer Corporation in Pomona, California, to aid its new orbital imaging company, which was expected to germinate after the launch of the experimental Seastar satellite. With the announcement, David Thompson also made his long-range plans clear: "The satellites we build will fly on our rockets; the sensors we build will fly on our satellites; and our ground stations will control the whole circus whizzing around above the Earth."

If the goal was to build a corporation with a lean vertical structure—one that suited the founders' vision and goals—quickly and efficiently, then mergers and acquisitions provided an attractive gambit. Few people in the corporation knew that as 1994 began, Orbital's executives also had initiated talks with executives in several long-standing and respected aerospace companies, more than one of which would result in another significant merger before the year was out.

By 1994 many of the historical alignments within aerospace—in satellite and rocket industries and inside venerated NASA—fell out of order. Traditional relationships suffered, interests shifted.

"It's a billion-dollar game with hastily arranged alliances," said one Wall Street analyst of the grab for orbital positions in the new LEO frontier. Terms like *interoperability* and *proprietary information appliance*, *hypertext transfer protocol* and *personal communication services* entered the aerospace lexicon at tongue-thwacking speed, echoing the new argot of terrestrial communications, which was quickly merging industries in cable, wire, wireless, and vastly expanding telecommunications networks across the globe. Anxieties in aerospace, an industry that had become strangely, quickly outmoded—"good old boys' clubs for retired air force joyboys," according to one industry critic—ran high.

"We see all these 1,600-pound gorillas coming after us," reported Larry Yermak, of Fairchild Space and Defense, a Maryland company with a great heritage in aerospace, speaking at a satellite conference in Washington in April '93. "A lot of the romance has gone out of this business. Right now, we're talking about basic survival."

The changing nature of the business also created peculiar deformities in revered pioneering companies. COMSAT, for instance, the historic satellite communications company, announced it was seeking a National Hockey League franchise as part of its developing interests in the entertainment industry, which already included a National Basketball Association team, the Denver Nuggets. "We stopped being a satellite company several years ago," COMSAT's CEO Bruce Crockett told a reporter at *The Baltimore Sun.*

INTELSAT, the international consortium that operated a kind of monopolistic sovereignty worldwide with 400 companies and 20 satellites, suddenly felt the sting of competition and announced that it would become more of a "commercial entity." The competitor? An unlikely entrepreneur from Connecticut, René Anselmo, a private citizen who had purchased and launched his own satellite in 1984 called PanAmSat. Prepared to go toe to toe against INTELSAT, Anselmo's business began delivering quality communications services to Latin America and undermined Intelsat's credibility in the global marketplace. In the gentlemanly sphere of Intelsat's monopoly business, the tenacious Anselmo chose to flaunt as his company logo a pissing dog underlined with the motto: "Truth and Technology Will Triumph Over Bullshit and Bureaucracy." By 1994 he had built a billion-dollar communications business with satellite services multiplying rapidly across the globe.

An explosion of a new technology in small aperture satellite antennas (VSATs) had created an enormous tide of new business for commercial satellite services. Notorious media mogul Rupert Murdoch paid $525 million for a satellite service based in Hong Kong called Star TV and then turned to the developing world as a lucrative market in which he could sell American mass culture. Delivering television reruns of *The Simpsons, Baywatch*, and *Oprah Winfrey*, as well as up-to-the-minute news reports from the Cable News Network, to large populations in China, the influx of satellite technology caused Communist leaders to declare the use of satellite dishes illegal, but then, according to a story in *The New York Times*, they failed to enforce the law because they feared rampant, potentially violent protests.

But the popularity of satellite services was not just restricted to China. In March 1994, Saudi Arabian officials banned satellite dishes as "un-Islamic," ordering more than 150,000 dishes dismantled; and Hindu fundamentalists in India went into a rage over an Asian version of MTV. In Egypt the governor of one province complained that the furniture industry in his district was dying because "the workers are stuck in front of the sex scenes coming off the dish."

As satellite dishes became smaller, more powerful, and less expensive, Third World governments realized they could not stop the flood—satellite antenna kits could be smuggled easily across borders and pieced together without any expertise or technical knowledge. In Iran alone an estimated 50,000 satellite dishes covered rooftops across the city of Tehran, and television reruns of *Dynasty* provided the city's most sought-after entertainment.

At the same time, an unexpected efflorescence of a certain kind of satellite service brought unique and inexpensive services to mass markets—the kind that Orbital hoped to capture someday. The most commercially viable new technology came from the most astonishing source of all, the Department of Defense.

Since the launch of Sputnik, American military planners had nurtured an interest in using the Doppler shift of radio waves from space to aid navigation on the ground, and by the 1980s the U.S. Navy and Air Force had combined technical talents to launch a constellation of satellites in low-Earth orbit for position determination of ships and aircraft. Comprising twenty-four satellites, the Global Positioning System created a strategic resource that proved invaluable in the Gulf War of 1991.

Interestingly, DOD had reckoned on the system being of commercial value someday and expected to offer GPS services to the public. But once the system went into effect, the enormous interest from boaters, surveyors, farmers, even backpackers was so astonishing that a number of new private ventures sprang up overnight to take advantage of an opportunity to use military hardware in space.

The market for GPS receivers boomed in the first years of the decade. As commercial outfits found ways to get around the military's effort to purposely degrade the system for general public use, sales began to skyrocket, prices drifted down to the $500–$600 range, and manufacturers predicted the cost for a GPS device would meet mass-market standards and plummet by the end of the 1990s to $100 or less.

More than any other satellite service, GPS underscored David Thompson's notion that with the advent of microelectronics and the rediscovery of the low-Earth frontier, the benefits of space would, as he put it, "come down to earth." The consumer market for space-based services—naturally global in scope—offered just the kind a gold mine he had predicted.

Never ones to miss an opportunity, as of March of '94, Orbital's executives had explored the field of companies in the United States selling GPS receivers, looking for yet another acquisition candidate. They imagined that a merger would give Orbcomm and their little Orby handhelds an ancillary feature with remarkably broad appeal. Time was running short, though, and there was no longer comfort to be taken by underestimating the competition.

With a few tricks still left to play, the Orbital Sciences Corporation needed nothing more than to launch its Orbcomm satellites as soon as possible. The rest of the world would soon realize that hype about a billion-dollar race to settle the low-Earth frontier was neither as absurd nor as grandiose as it once sounded.

Investors and service providers like Santiago Aguerrevere realized they would have more choices in only a few years. If Orbital wasn't quick, the competition would steal away with their customer base, ravish their markets, and leave the Orbcomm business to rot like a broken-down Conestoga on the prairie.

That night in downtown Arcade, bikers raced their Harleys in and out of the trailer park, rumbling down Main Street, it seemed, for hours. Ames Department Store saw a rush by locals shopping for spring clothes. At Josie's bar, a heavy metal band named Fire and Ice played late, awakening customers at the motel next door with a particularly noxious version of "Stairway to Heaven" at 4 A.M.

The engineers from Orbcomm noticed nothing. They had driven back to the motel late at night and slept a few hours in states of deep concentration. At six that morning Stan had met the Venezuelans stirring around the coffeepot in Josie's lobby and taken them out to his chosen breakfast spot, the local Putt-Putt. Then he drove them back up the mountain, where he could track satellites and test the Earth station antenna in the same state of focused devotion as he had for days.

After a lunch of beans and rice, Aguerrevere talked about the site preparation for Venezuela's own Earth station in the mountains 20 miles south of Caracas. His company had already made the down payment on land and would soon begin getting permits, clearing trees, building roads, and laying power lines.

"We have to show pictures to our customers," he explained, as his colleague continued snapping photographs around the blockhouse that day. "We have to show them you are real."

The first Earth station tests in Venezuela were already planned, too, set for September, and all his clients expected the service to start by December.

"Once people know this system exists," he said, "they will sign up quickly."

None of the Orbcomm engineers corrected him.

September? December?

By five o'clock darkness fell with a quiet snowfall, ushering in another lonely winter night. Pink shadows of passing clouds painted the snow, angling out along the muddy path to the road, mixing with the blue sheen of ice sloping down in all directions. Lit with bright lights, the radome glowed, and inside the Earth station engineers continued to track dozens of satellites whipping ceaselessly across the horizon.

"We got two more," Mazur said. "Which one you want?"

"Fielder's choice," Figueroa murmured.

Gathered around their computers, they knew nothing about the snowfall outside or the fall of darkness. They were much too busy.

A fog settled in on the mountaintop, and the gentlemen from Caracas drew their chairs up to the consoles, increasingly eager to learn how to track and send signals to little beacons that tumbled between them and the stars.

CHAPTER FIFTEEN

March 1994

Interference

After the third meeting of the day about Grace Chang's ill-fated Earth sensor, Dave Steffy went back to his office and began stuffing everything from his in-basket into his briefcase—a sporadic task he called "vacuuming the in-bin."

He reached into a filing cabinet to remove a stack of academic papers about antennas to take home for bedtime reading. At least for one evening in March, he expected to get home before his kids went to bed. With his briefcase filled, Steffy imagined he would knock off the few last items on the day's "to do" list while he watched the NCAA basketball tournament.

When he turned to leave, Mark Krebs walked by the doorway.

The engineer's eyes looked milky, shot with red streaks like boiled shrimp, the result of too many consecutive days writing algorithms on MatLab. He had not been happy about the outcome of their meetings about the Earth sensor.

To save money, Steffy had vetoed his request to install a new set of magnesium brackets that would have improved the pointing accuracy of the attitude-control system by one degree. Still anxious to meet specifications, Krebs needed every bit of margin he could scratch up.

"Remember, Mark, all these things have to do is produce revenue," Steffy said, poking his head out the door. "That's the most important thing."

Krebs spun around, and kept walking away backward. "Oh, okay, 'enhance revenue'—well, that just about covers all my worries," he sniffed.

"What—more or less worried?" Steffy asked.

Krebs stopped, whirled around, and held up the Earth sensor, a silver object that he palmed like a hand grenade.

"See this?" Krebs said. "There's no margin in this thing. If there's an answer anywhere, I still think we should spend the money. You can say this is a fixed contract and a spec's a spec, but I haven't figured out yet how to change the laws of physics."

Krebs turned and kept going, straight to the lab, his home away from home.

Just then Tony Robinson came up the hallway.

"Don't worry, I'll build Krebsy a shim," Tony said. "'Course, that'll cost you money just like a new one."

"I expected you to come through," Steffy said.

Something about the comment brought a strained look to Tony's face.

He backed his boss into the office and shut the door behind them. Steffy seemed puzzled. He didn't remember the staff meeting that afternoon, when he had blamed the Earth sensor problem on an alignment error by Tony.

"We're doing new stuff here," Tony said. "If you could get everybody to quit redesigning things, we could freeze the layout and avoid problems like this."

In fact, the layout continued to change regularly as engineers fiddled endlessly with their designs. In the past week, for instance, after new test data came in from an antenna range, Tony had been forced to expand the trough inside the satellite by two inches, a dramatic development for a spacecraft already squeezed to the gills.

"And don't forget you took away my best designer," he continued, a topic Tony never seemed to purge. "Then you stick me with a bunch of inexperienced rocket designers who don't know a satellite is a precision instrument."

"But it's always like this, Tony," Steffy said. "You never find problems until you start to pack models in the lab. That's what they're for."

"You might as well hire 'em from York Heating and Air Conditioning!" Tony exclaimed. "These new designers don't know anything about small spacecraft. But as soon as there's a problem—boom!—I get blamed."

"It's no different than what the software team's going through," Steffy said.

"Well, with the next satellites it better be different," Tony said.

"All our jobs are gonna be different," Steffy said.

"Believe me," Tony said, "it's the first thing anybody does when something won't fit—they look at me. 'You're the mechanical engineer—how come it don't fit?'"

"I agree with you—"

"Then get off my back!" Tony said hotly.

In the past month redesigns had sent Tony back to configuration trades on the power system and attitude control. He was moving boxes, looking at inertial ratios, moving the center of gravity. Power margins kept changing. He was trying desperately to damp out the frequency on the solar panels when they sprang open. Basic ground-level analysis, at this late stage, kept him hopping.

"I'm not complaining," Tony said, "because it's all going very well so far. But I wish you'd let people know how good we are instead of criticizing me and my guys in front of everybody."

"I didn't mean to take anything away from you, Tony."

"Well, it's a catch-22," Tony said.

Steffy pulled out a chair at his conference table and sat down. Tony was one of the best mechanical engineers he had ever worked with. His inventive design for the antenna had probably saved the project. The trades he made to keep the weight of the satellite below 95 pounds, despite constant changes in designs and materials, was nothing less than spectacular.

"We're not in an engineering crisis, Tony," Steffy said.

Tony shook his head.

"Tell that to our wives and children," he said, sounding particularly bitter. He couldn't seem to back off.

"No, seriously," Steffy said.

"I am serious," Tony said. "You know, one of our best guys got an ultimatum from his wife yesterday: 'You don't come home at least one night this week, I may not be here when you come home this weekend.'"

Steffy nodded. He had heard stories like that. The same tensions had crept into the marriages of Pegasus engineers when they neared the final months before their first launch. He started to say something about Pegasus, but Tony stopped him. Tony's wife was pregnant, at home with a three-year-old. He didn't need another pep talk.

"Look, here's something you can control," Tony said. "Every once in a while, you should say, 'Okay, Dan, Rob, Eric—here's some money. Take your wife out. Take her to a nice restaurant. Go out on the town. This is the Orbital way.' You get the wives to buy into the program, you'll—"

"That's dangerously close to being sexist," Steffy said.

"But it is the wives—okay, spouses—*families* who suffer the most," Tony continued. "If we're looking at another eight or nine months working at this level, a gift of 50 or 100 bucks once in a while would go a long way to taking some of the heat off at home."

Steffy's face wrinkled with concern. His family suffered, too.

If only it could be so simple. Many days he wanted to tell everybody on the team, "Hey, I don't want to see any of you in here next weekend." But there was never an opening. Either they were climbing out of a miserable week and building momentum into the weekend, or they sensed doom and no one wanted to relent.

So instead of responding to Tony's suggestion, Steffy rambled on about the latest launch schedule, the risks and problems with the Qual and Flight spacecraft, and soon he was lost in a forest of detail, the place he sometimes found himself at the end of an exhausting day.

Tony listened for a while, then gave up.

"Okay, at least think about people's morale," Tony said, finding a moment to intercede and back out of the conversation. "And don't forget mechanical—that goes straight to my bonus."

When Tony left, Steffy pushed his chair up to his desk and examined the latest financial spreadsheets.

There was just no way he could find Mark Krebs money for a new bracket.

For the first time Morgan Jones looked bored in collegiate uniform—leather sandals, cotton shorts, wrinkled T-shirt—as he shepherded RF tests solo in Boulder's large, lonely lab.

Backsliding into the town where he had graduated just two years before, a thousand miles away from the Freshout team in Virginia, Morgan wore an expression of utter ennui that explained everything.

At the start of the previous summer, while still living in Virginia, he had risked his job by threatening to quit if the company refused to move him back to Boulder to be with his college girlfriend. He had won the gamble, then lost the girlfriend. Within two weeks of arriving at Jan King's Colorado mountain satellite paradise, his honey dumped him, and he spent his time playing catch-up for Boulder's senior engineers. No longer the OSX guru or the one always in demand, instead Morgan aided more senior engineers whose reputations as the project's prima donnas overshadowed his status as lackey . . . assistant . . . gofer . . . novice . . . Freshout.

If the project needed prima donnas, it was in Boulder, where the satellite's transmitter and receiver were being made. Morgan now worked in their shadows.

The transmitter rested on the blue workbench next to him, about stomach high, connected by a serial line to a development board the size of a small cafeteria tray. The green monitor on the spectrum analyzer, perched at his left, displayed a ragged blaze of jittery impulses, a scorched path of bit packets streaming across the 137 MHz band. A moldy growth of fly wires and hairy jumpers spread across the top of the transmitter where Gordon Hardman, the senior engineer for RF communications, had corrected flaws on his Qual board.

Gordon had taken 16-gauge Teflon-coated copper wire, turned twelve times around a mandrel on a drill bit, and applied it to his high power amplifier, hoping to stem an ongoing battle against electronic spurs. Dave Steffy joked that the transmitter looked "like something you'd see in one of those cheap science fiction movies when somebody powers up the Death Ray." But Morgan knew the transmitter would never pass Qual tests in that kind of shape, and it would most certainly be downgraded to EDU status. The Death Ray had become Morgan's job to repair.

He reached out with his left hand to the analyzer and slowly spun a gray dial on its face. Then with his right hand he punched a few buttons to zero in at 137.52 MHz.

A test set, built by Torrey Sciences in California, drove a weak signal into the transmitter coupled with interference from a noise generator. Morgan looked puzzled as the blaze of bits changed course. He spun the gray dial again, popped the buttons, checked to see if he had nailed 137.52, and went through the test again. Then he tried it a third time. And a fourth.

On each attempt Morgan saw the signal-to-noise ratio escalate unexpectedly nearly ten times beyond spec, while the flow of gray numbers on the blackened screen of his PC monitor kept burping up a bit error rate of zero.

Couldn't be, he thought. This is absurd. Errors should be piling up like mangled engine parts in a junkyard.

Morgan peeked between the shelves of his lab bench across the broad room to an empty glassed office. Gordon was long gone.

It was hard to admit to himself, much less anyone else, but Morgan was in over his head. He had written code for the receiver and transmitter, true enough, but when Gordon started passing off actual hardware—the transmitter itself—he hadn't a clue. Morgan's training had never provided him with a rare dip into the netherworld of spectral wizardry, a discipline that even Dave Steffy referred to as "black magic."

Sometimes when he had no other place to turn, he would look for Gordon. A tall, thin, dignified white South African, Gordon would peer down at their dilemmas, spread out in pieces along the bench, and in his finely tuned English accent, remark, "Oh, that's rather peculiar," or "Now that might pose a problem," and then disappear, leaving no explanation, no answers, no hope of insight. Sometimes he would be gone for days on end.

The fact was, since he moved to Boulder, Morgan had seen the influence of corporate politics at the engineering level. From the East Coast, when Steffy would call and ask about their progress, Morgan heard the Boulder engineers respond, often sounding agreeable, if not sanguine. But try to pin them down on a date when their EDU hardware would be completed or when this test or that would begin, and they would digress, elude interrogation, refuse to commit. Between pauses around the conference phone, Morgan would see them grimace as the program manager demanded target dates. When Steffy spoke of urgencies, they would silently shake their heads. As they sidestepped questioning, Morgan, too, remained silent. Everyone had a reason—technical and personal—to evade command.

By 1994 Jan King's satellite engineers felt they had endured quite enough. At one time hoping to run the Orbcomm enterprise himself, King had finally given up any notion that he or his engineers would win a piece of the long-term business. They knew Alan Parker was saying behind their backs that he wished David Thompson would close down their entire enterprise. But then they didn't need to hear him say it directly. They had more vivid evidence.

Over the year, this is what the Boulder crew noticed: Despite managing to snag a rare ride on the SR-71 at no cost during the fall of 1992, they never heard a word of thanks from Orbcomm; the excellent job the team had done building and launching the two OXP satellites went unrewarded; expectations that an ongoing analysis of the satellite links would become Boulder's responsibility were never met; offers to build the first test models of the Orby terminals were quickly rejected. Finally, out of frustration, Jan King sent Alan Parker a bid of half a million dollars to conduct a complete link analysis that would satisfy any questions about interference on the ground and in space in the Orbcomm bands. Parker responded angrily, claiming the estimates were outrageously high and deemed the bid an insult. At that point, their relationship—what little remained of it—collapsed completely.

By the spring of 1994, it was no wonder that the engineers in Boulder appeared to have lost interest in the project altogether. They would say privately that executives at Orbital made it clear they would be punished indefinitely for the loss of Orbcomm-X. Even though they shouldered responsibility for the most critical part of the satellite system—the valuable communications boxes—they did their work grudgingly at times. "There's a lot of frustration out here," one of the senior engineers told me one day. "It manifests itself in a simple way. Had we been able to work collegially with Orbcomm, Alan Parker would have gotten a tremendous value for his money. But after the way we've been treated, do you think we're going to work weekends and bust our butts or care very much when he calls? Okay, maybe we'll lose a little bonus money because we don't meet schedule. On the other hand, if you look at the value of Orbital's bonuses in terms of hours, we could make more money flipping burgers. We actually calculated it once—it comes out to something like 45 cents an hour. So when they call for work on Orbcomm, the question we ask now is, 'What's in it for me?' No more sixteen-hour days. No more freebies. Our people will do a little more jogging in the summer. We'll take vacations when we need them."

The Boulder division had changed into a far different place from the one it promised to be only sixteen months before, when Dave Steffy and Jan King eagerly joined engineers in the parking lot to tune the first cut of a satellite antenna. Once enthusiastic members of the Orbcomm team, Boulder's engineers now watched their piece of the business dwindle. After the launch of the first two Orbcomm satellites, they had no more work, period. Alan Parker didn't want them; David Thompson didn't intercede. By all indications, unless the Boulder office drummed up its own projects in the next year, Jan King's labs and offices would likely close for good.

While Morgan continued to fiddle with his equipment, another engineer, an older man carrying a satchel of plastic graphs and charts, stepped up to the other end of the workbench and flipped through his own paperwork. Morgan glanced in his direction, but didn't catch his eye.

"Lunchtime," Morgan said.

The man at the end of the bench didn't respond.

Morgan jotted down a few notations in a brown lab book and then, without another word, left for an early lunch—and a long one—from which he would not return for several hours.

The man who had sidled up on the bench was Mike Carpenter, a senior engineer responsible for building perhaps the most difficult piece of technology on the satellite, the subscriber receiver, the box responsible for accepting messages from the Earth and preparing it for retransmission. He had taken a battering over the past year, seeing his work undermined by faulty microchips, abrasive corporate politics, and an especially maddening set of specs that put his performance on the verge of theoretical performance.

The picture of him that afternoon in the lab moving up and down the bench, switching on and off pulses, commanding an entire sequence of simulations—driving bursts from an Orby terminal to a cruising satellite, and into Gordon's transmitter—reflected a carefully defined process, one that had besieged his life for months.

Outwardly calm and self-assured, Mike had become increasingly agitated by the project. The technical challenges still excited him, he would say, but constant deadline pressures, which he interpreted as exaggerated and manipulative, often turned the pleasures of engineering into a nerve-frazzling, everyday annoyance.

Interference of a different sort also frazzled his receiver. With the uplink band at 148–150 MHz, Orbcomm's slice of the electromagnetic spectrum exposed itself across two full megahertz to a range of unseemly and raucous carriers, some authorized, many not, doing everything from directing a top-secret Russian radar system to handling mobile radio traffic in the Pacific Rim. In order to find Orby's little five-watt signals as they fluttered up randomly from the Earth through a storm of blustery channels, Mike's receiver first would have to welcome every earthly signal into its front end, unfiltered.

Oddly enough, he had built the receiver to admit the full host of the world's interferers into his box. In a way the satellite's subscriber receiver would have to act more like a military electronic countermeasures receiver—a device immune to jamming from foreign sources—than like a traditional radio receiver operating in a noise-limited environment. The only significant question was, if he admitted in every signal across the band, at what point would all those signals strike a convincing assault against the satellite and shut it down?

Mike stopped for a moment to clean his glasses.

His orderliness had created an impeccable standard. The circuitry on his board, unlike Gordon's, contained no fly wires. The plan he had devised to pluck Orby's little signals out of the terrestrial storm showed itself visibly, in miniature, neatly spread sequentially across the physical dimensions of his receiver. There he had created a clear path with a field of micro, digital electronics.

Standing at the bench, Mike retraced the scheme he imagined would make his receiver most adept at sorting through the noisiest channels in the spectrum. Comprised of two assemblies, with two circuit boards in each assembly, the receiver eventually would fit inside a single, hollowed, multilevel aluminum housing that would hold the boards on separate levels. Gold in color, the housing had a primitive look, like a model of an Aztec dwelling. At the moment, however, critical pieces of hardware had been released from their housing, and he had scattered them separately across the bench, where he could keep them visible and accessible to his touch.

Across one full side of the board stood raised banks of seven different receivers—six matchbook-size demodulators that would handle Orby signals from the ground and one that would act as a scanner, constantly determining how much energy existed in each of 820 channels across the 2 MHz band in space.

Relying on a statistical algorithm, the scanner would report its predictions of the clearest channels, selecting six considered most likely to be open for a range of five seconds. Once the algorithm identified the six channels, the satellite would send a signal back to the little Orby on the ground assigning it a channel and, simultaneously, alerting each of the six satellite receivers to prepare to snag a message firing back up on that same channel from the Orby.

From space the subscriber antenna would sponge up all signals across the band and flush every one into a low-noise amplifier. The reason for allowing them all in was simple: Orby's signal would enter the satellite at an average power of minus 112 dB/m and, at its weakest link, minus 125 dB/m.

Rather than build the receiver to block out the power of other radio signals in Orby's uplink channels, Mike had to accept the entire onslaught or risk losing Orbcomm's signal completely. In other words, before he could find the weak Orby signal, he had to open the gate and let everything in.

Dave Steffy once compared Mike's task to the effort of someone at the goal line of a football field trying to hear a message whispered at the opposing goal line during a noisy half-time show while someone screamed into one ear and blasted a foghorn into the other.

A more technical explanation went something like this:

In satellite communications signal strength breaks down into sets of numbers called decibels, or dBs. The reason engineers use dBs instead of ordinary numbers is quickly apparent. For example, although the signal transmitted from an Orby from the ground would look in space just as it did when it left the ground (the wavelength would be the same), its signal power would be, naturally, much less, a change in strength called *attenuation.* Like yelling to someone across a crowded football field, where sound grows weaker over distance, a radio wave will grow weaker as it passes from its ground antenna into space. In the case of the Orbcomm system, signals generated by a little five-watt Orby antenna were estimated to be one million billion times weaker by the time they reached the Orbcomm satellites, 785 kilometers overhead.

One million billion times less than its initial power: In the lingo of RF engineering, the figure can be stated in decibels, a logarithmic expression that describes a power ratio—with Orbcomm the average signal would arrive within earshot of the satellite antenna at a power of minus 112 dBs. The link budget for the system showed that the satellite would have to distinguish Orby's little signal from every other radio wave passing its way at minus 112 dBs, on average, and at its weakest, minus 125 dBs, a level so low that even the spectrum analyzers used in Orbital's electronics lab could rarely detect it during simulations.

To make it possible for the receiver to pick out such a rarefied signal, Mike first employed a low gain device on his receiver that allowed all the racket in the band into the satellite, but at such a low level that it would not overload the receiver. Then those same signals slipped discretely over to a frequency oscillator.

The oscillator converted everything in the band from 148–150 MHz to 10 MHz, the downconversion making it easier later to amplify the signal again without overloading the system. The signals then split up six different ways and entered the six demodulators, which finally searched for and identified the one channel assigned to the Orby back on Earth.

At that point a filter would, for the first time, effectively shut out all the residual junk that had entered the antenna initially. The Orbcomm signal would then be amplified by a factor of 20 and run through an analogue to digital converter, which, as the name implies, translates the analogue signal into a digital format. The signals would convert to baseband in the digital format and pass on to six digital signal processors.

"It's all math after that," Mike explained that day as I watched him work. "We're unsmearing the data" at last, sending the Orby messages along to the satellite transmitter to be returned—loud and clear—at a lower frequency, back to the Earth.

It was Mike's misfortune that Orbcomm Inc. had specified the most rigorous signal-to-noise ratio for his receiver. Without data from Orbcomm-X, David Schoen claimed that no one in the world—other than the U.S. and, possibly, Soviet military—really knew the actual level of interference in the 148–150 MHz band. He played the links conservatively.

How conservatively? If the Canadian paging system represented the world's largest interferer in the 148–150 band, Schoen's specs suggested that every place on Earth had its own Canadian paging system running full blast every minute of every day. His choice was ferociously, perhaps impossibly, conservative.

Fierce arguments about Schoen's so-called "interference rejection specs" had dragged on for the entire two years of the Orbcomm project, inflaming tempers and adding to the feelings of alienation within Jan King's group. Steffy had fought Schoen aggressively in meetings with the company's lawyers, and despite the lawyers' insistence he had still refused to sign off on the specifications.

Some people believed the specs were set at an extreme limit for no reason but to make sure David Schoen had covered his bets. Since the beginning it had been Schoen's job to manage the entire communications network for Orbcomm—its terrestrial routing schemes, the flow of messages from the network control center to the gateway Earth stations, through the satellites and from the Orby terminals. He was, in a sense, the system's technical creator, the King of Technic. Importantly, it was also his responsibility to write the algorithm for the scanner, that seventh receiver on Mike's board, using a new technology that Jan King, in fact, had created, called the Dynamic Channel Activity Assignment System (DCAAS, pronounced "D-Cass").

Without a real picture of all 820 channels in the 148–150 MHz band, Schoen realized he could do nothing less than anticipate the worst possible levels of interference—everywhere, at any given moment. So with unyielding authority he had written the DCAAS algorithm accordingly and passed the specification on to Dave Steffy, Jan King, Gordon Hardman, and of course Mike Carpenter.

Now, at every junction on his receiver where cables ran in from the test set or from his PC or from Gordon's transmitter, Mike could see rollicking interferers simulated to race like a thundering herd of wild horses into his box: thermal noise louder than the Orby signal itself; bombastic charges leaking from Gordon's transmitter; huge interferers like the Canadian pagers; hundreds of unidentifiable, weak signals at minus 52 dB suddenly adding up that would threaten to overload his system. Contamination, leaks, distortions of phase and amplitude, wildly out-of-control intermodulation products, an entire cascading river of radio waves crashing in, beating down, jazzing up, splitting, combining, reducing—each burst taking less than 500 milliseconds to enter the satellite and return to the Earth.

He watched the action from the analyzer's green monitor, then leaned over to read the outcome, displayed as the bit error rate, on his PC.

His test had run all night so the amount of data was overwhelming. He stepped back to the spectrum analyzer and watched evidence of Orby bursts squeezing signals into his system, then followed them through a series of peaks that rose and fell across his screen.

Mike reached out and touched his monitor, counting the number of peaks—each tip higher than the one before—until he counted seven at the highest point, and back down, seven again, until the signal fell off into a horizontal blur, pictured at the edge of his screen like short grass at the base of a mountainside. Tapping a button, he called for a printout, then another, and another.

The paperwork, graphed in black and white, indicated that the problem he had fought over the past six months was now nearly resolved. He had, at last, beaten back the lion's share of interference in the band. Not perfect, but nearly so.

Mike grabbed his charts and packed his satchel.

He wasn't expected in Virginia until the next morning, but he had anticipated using the extra hours there to adapt his story to the rest of the team's. The last time he had conducted a demonstration for Teleglobe's consultants, a chip in his board had burned out and the test failed, a miserable and egregious error that, despite it not being his fault, left him more cautious before formal presentations.

In a recent missive to Steffy, Teleglobe's consultants had written that they believed only three issues remained unresolved on the project: noise environments on the Earth; the launch schedule date, which they believed was at least nine months off target; and the satellite's communications system, particularly the receiver's proficiency against interferers.

Mike Carpenter felt intense pressure but also some ambivalence about his future in the corporation. As he left the lab and drove to the airport, he probably knew that his job at Orbital was in jeopardy, regardless.

And he was still 3 dB short of the goal, behind by a factor of two.

David Thompson might not be willing to yield, but at this stage of the project, maybe that didn't matter so much anymore.

Bulletin Board

"The only incentive we get is this 'We gotta-do-it-for-the-team' crap. But after a while the pressure gets to you, you just get numb to it. Nothing they say matters anymore. You work as hard as you can, and after so long you just have to say, 'I'm outta here.'"

—Steve Gurney
Satellite engineer
March 1994

GREGG BURGESS: David Thompson says he's serious about making this schedule, but I don't see any action. At Hughes we could always bring in a few extra people to go kill a problem.

ERIC COPELAND: Did someone say kill? That would work nicely.

—Staff Meeting
March 1994

"Most risks associated with Orbital's stock concern the typical technical, financial, and regulatory matters involved in the satellite business. In the Orbital prospectus, Lehman Brothers offers potential stock buyers some key milestones to watch:

- the successful launch of Orbcomm satellites in 1994;
- the initiation of intermittent Orbcomm mobile communications service;
- receipt of Orbcomm full operating licenses from the FCC;
- completion of the final financing agreement with Teleglobe."

—*Satellite News*
Underlined on Dave Steffy's door
March 1994

I see the stock going to $50 within the next 18 months.

—Gary J. Reich, Prudential Securities
Barron's interview
March 21, 1994

MORGAN JONES: This is weird. You work for years on something, snarling at people and pissing them off, and one day—bing!—it works! And for five minutes you feel elated, and then the next moment you're wondering what else there is to do.

ROB DENTON: Get back to work, Morgan.

MORGAN JONES: Fuck off, I want to enjoy my moment. For one day, at least.

ROB DENTON: Maybe you better schedule some vacation before we find some other crisis for you to work on.

MORGAN JONES: Shut up and let me bask in victory.

—Lab conversation
April 1994

"Does Orbcomm sound too good to be true? It may be—at least for a while. The system is at least a year behind schedule, although the first two satellites should be in orbit by the time you read this article."

—*Popular Mechanics*
April 1994

"Let's just hope they don't want to see a satellite at the board of director's meeting."

Dave Steffy
Orbcomm satellite program manager
April 1994

"Every day seems like Monday to me anymore. I came in here on Friday and it was like Saturday. Nobody was here but me and Rob. It was just empty because Steffy was on vacation and everybody else is so burned out. Then I was here yesterday and it was like a ghost town and then today. I really don't even remember where I am anymore. "

Eric Copeland
Satellite engineer
May 1994

"There's probably a big biz in LEOs, but I don't think there's a big business for lots of LEO companies. There will have to be some sort of decision on what's the best technology."

George Gilder
Upside
June 1994

Proposed LEO systems, 1994

Aries, Constellation Communications, Herndon, Va.:	48 satellites
Ellipso, Ellipsat Intl. Inc., Washington, D.C.:	14–18 satellites
Iridium, Iridium Inc., Washington, DC:	66 satellites
LEO One,. Leon One Panamericana, Mexico City:	1–36 satellites
Odyssey, TRW Space & Electronics Group, Calif.:	12 satellites
Orbcomm, Orbital Sciences Corporation, Dulles, Va.:	36 satellites
Signal, the Russian Republic:	48 satellites
Starnet, Starsys Global Positioning, Lanham, Md.:	24 satellites
Taos, French Space Agency, Paris, France:	12 satellites
Teledesic, Teledesic Corp., Kirkland, Wash.:	840 satellites
VITA, Surrey Satellite Technology, UK:	2 satellites

Most Common Spacecraft Errors

Payload:	39 percent
Attitude-control system:	20 percent
Reaction control system:	9.6 percent
Electrical power system:	9.3 percent
Mechanisms:	9.0 percent
Command and telemetry:	6.8 percent
Thermal control:	3.9 percent

Corollaries to Murphy's laws

by Edsel Murphy

A. The necessity of making a major design change increases as the fabrication of the system approaches completion.

B. Firmness of delivery dates is inversely proportional to the tightness of the schedule.

C. Dimensions will always be expressed in the least usable terms.

D. Original drawings will be mangled by the copying machine.

E. All constants are variable.

F. In a complex calculation, one factor from the numerator will always move to the denominator.

G. Identical units tested under identical conditions will not be identical in the field.

H. If a project requires *n* components, there will be 'n – 1' units in stock.

I. A dropped tool will land where it can do the most damage. (Also known as the law of selective gravitation)
J. A device selected at random from a group having 99% reliability will be a member of the 1% group.
K. Interchangeable parts won't.
L. After the last of 16 mounting screws has been removed from an access cover, it will be discovered that the wrong access cover has been removed.
M. After an access cover has been secured by 16 hold-down screws, it will be discovered that the gasket has been omitted.

—Found on satellite team bulletin board
June 1994

"Orbital has been trading in the low $20s since late March, when the stock hit a 52-week high of 26. To get the stock moving up again, analysts said the company must successfully launch the first of its Orbcomm global communications satellites."

—*Aviation Week and Space Technology*
June 1994

"Maybe if I leave this job I should just go somewhere totally different. Maybe I should get out of aerospace. I don't know. It's weird. You remember how they were going to tell us when we'd be laid off and give us a few months notice and all that? Well, that's all gone now. We just get laid off. No notice. 'Sorry, you're out of here.' It wasn't even announced or anything. You just find out like word of mouth. Everybody's sitting around here now wondering when the layoffs are coming, and now it's like, forget it. When your day comes, you'll get a phone call, and you pack your bags right then."

Steve Gurney
Satellite engineer, July 1994

CHAPTER SIXTEEN

August/September 1994

Old. Boring. Really Depressing.

The season's most oppressive heat settled in like a steam bath by mid-morning and clung moistly to his clothes. Morgan plodded along the winding concrete walkway linking the new corporate office building to the million-dollar high bay back in the woods, feeling happy and, for once, self-satisfied.

The late summer weather of Fairfax County seemed familiar, reminding him of one of the little-known facts he had learned while living in D.C. the previous summer: because of the region's excessive heat and humidity, British soldiers dispatched to the Potomac swamps during the Revolutionary War received compensatory pay for "hazardous duty."

Why Orbital had waited so long to pay compensatory wages to its over-worked Freshouts, he would never understand. But now, in August 1994, more than two years after he started the project, like a good soldier, Morgan felt obliged to finish his commitments even though he was all but finished as an Orbital engineer. At the same time he anticipated one final surge in his paycheck.

He had flown back to northern Virginia for the last time in early August, hoping to complete a series of performance runs on FM-1 and FM-2, the two satellite flight vehicles. Even before he arrived, he imagined walking into Steffy's office and giving two weeks' notice. Now that he had done just that, he felt lighter, confident that his decision was correct. After this last round of testing, he would go home to Boulder and start his own software business. Given the new incentive program, he also expected to collect perhaps a few thousand dollars, a bonus that would elevate his final experience at Orbital into a potentially celebratory occasion.

In fact, he felt a little sorry to go. His deepest disappointment had less to do with Orbital than with aerospace in general. Given the long molting season of the industry, he imagined he might as well start his own company in Boulder, an enterprise completely unrelated to space—an Internet support service—as the future of aerospace appeared grim for bright, independent young engineers like himself, and prospects for a fulfilling career in aerospace-land seemed less likely than prosperity and reward on the cutting edge of software development.

The thoughts of woolly Brits and hazardous conditions slipped from his mind as he climbed the steps to the parking lot in front of the high bay. Deadening heat; one final week under a fluorescent glaze that masked day and night; twelve-hour trials pinpointing software errors and crushing bugs; pizza dinners, stale coffee; one final reward, long overdue—a small, financial windfall.

Steve Gurney slipped off his tennis shoes and socks and massaged his feet. For once, he had arrived home at a decent hour, a little after six.

Last night's dinner bowl sat next to him on the table beside the sofa. The leftover ring of spaghetti sauce, now crusted, reminded him of other dishes afloat in a cool bath in the kitchen sink. He looked up at the ceiling and stared at the holes he'd drilled months ago in an aborted effort to replace the house's frayed 1920s wiring.

His eyes drifted over to dangling vines of ancient electrical wires wrapped in shredded cloth. Dark pocks in the ceiling still needed spackling. Despite an agreement with his landlord to paint the rooms, he never had found time to roll on the first coat.

After more than two and a half years, in fact, Gurney had never even found time to buy furniture, except for the couch and a few chairs. The room itself, one large high-ceilinged space, still looked only half lived-in, almost abandoned, spare but for clumps of scattered cast-off clothes and two enormous stereo speakers, which at the moment vibrated under the grind of an au courant band, Soundgarden.

He was thinking about moving. He was thinking about a summer vacation—to Germany or Turkey or, closer to home, to Nags Head. He was thinking about applying for graduate school. He was thinking that his résumé resided on a floppy disk at his parents' house in Bethesda, and he would have to go home and retrieve the disk over the weekend.

He picked up a sneaker, considered for a moment bicycling over to the Howard University campus for a pickup game of basketball, then tossed the shoe back on the floor. His knees hurt, anyway, from too much hoop. One night on the way home he'd been jumped and robbed, too, so he didn't feel perfectly safe pedaling into the city after dark—or maybe it wasn't that. In any case, no basketball tonight.

Sometimes he thought he was just getting old. Not too many years until he turned thirty. Restless and tired, his mind wandered.

A plague of broken objects had descended on his life that summer. The air conditioner in his apartment failed first, exposing him to Georgetown's abysmal heat and muggy nights. Then his Saab developed an oil leak, the chain on his new motorcycle snapped, and his friend Peter broke up with his girlfriend, leaving Peter with nowhere else to go but Gurney's apartment for an indefinite season of grief and self-immolation. He didn't mind Peter staying there, but at the moment it was nice to have the place to himself.

Dozing in his apartment late at night after work, doors and windows thrown open around his second-floor flat, cicadas whirring dizzily, tree frogs singing wearily in the heavy limbs that shaded his porch, Gurney had begun to dream about vague perils. Some nights he would awaken with his jaws aching, not remembering his dreams, feeling only an urge to take his new motorcycle out for a ride on the highway, to run through curves on I-66, to slant into the wind at 90 miles an hour.

Damp, dank Georgetown called Gurney out to play, and then he would blink his eyes open and remember his predicament. He expected to be laid off from work in a couple of months, possibly as soon as September.

The satellite's software had taken a ponderous course toward disrepair and self-destruction, which should have surprised no one, least of all Gurney himself, who, like a prophet in his own country, had never succeeded in getting his bosses to consider seriously his predictions of doom and disaster. With Orbcomm's launch date pushing further toward the end of 1994, several new software engineers had joined him and Annette on the team, finally allowing the satellite's operating system and its quirky incantations to undergo proper exhaustive scrutiny. Panic in the executive suites had resulted in the hire of the few experienced software hands, but by Gurney's way of thinking, good help had come too late.

He hadn't really wanted to leave Orbital, unlike some of his teammates who had recently taken GREs, applied to business school during the spring, and come into work lately announcing high scores. No doubt Eric Copeland would be at Stanford by the fall of 1995; Rob Denton talked about leaving to start his own online game business; Morgan Jones had mentioned a few ideas for a start-up. Gurney thought about each of them and reflected on his own plight.

For the most part, he lived simply, always kept his mind in the present. He liked it that way. No fear, no worries. For a long time, he thought work was, at worst, tolerable. But then he apparently had been less aware of his own submerged feelings of discontent than his colleagues.

One day, for instance, he went into the lab and heard everyone talking about priority lists. One, started by the man who had replaced Bob Lovell as chief executive of the Space Systems Division, contained names of specific engineers who would be laid off as soon as possible from July through December, in an effort to staunch financial losses in the division; another, kept by his friends on the team, predicted which ones among them would quit before the launch.

Gurney didn't know where he placed on the executive's list, but when he learned that his name appeared at the top of his colleagues's, he began to wonder if he had missed something. Did his attitude signify something important that he should pay more attention to?

Then when he went to see a doctor about the soreness in his jaw and learned that he was grinding his teeth at night over these mysterious anxieties, he thought, why should he hang on to the bitter end anyway? He wasn't unhappy with Orbital. Maybe he had just grown so accustomed to living with impending doom—the software's and possibly his own as an engineer—that he had overlooked one significant aspect to his lifestyle: the future.

The past six months had been particularly challenging. In March Bill Gates and Craig McCaw announced that they would extend their claim on the Internet by lifting it skyward, launching their own LEO satellite system, a $9 billion swarm of 840 satellites called Teledesic.

Gurney remembered the day the news broke, as the satellite engineers sat in a conference room eating lunch and a day-old copy of *USA Today* floated from hand to hand.

Eight hundred forty satellites!

"How many satellites did he say?" someone asked, as Mike Dobbles read aloud selections from the front-page report.

"Hey, look!" Dobbles said. "Alan Parker's quoted in here."

Sure enough, Parker was widely quoted as president of Orbcomm, still considered the front-runner in the race to settle claims on the LEO frontier.

("What's Bill calling this thing?" Parker had asked reporters when they phoned, "Telekinesis? Telekaniption?" He was clever, characteristically unflappable. "I thought it was Telegesic. Sounds like something you'd put on poison ivy, doesn't it? Or maybe it's a cough syrup. A pain reliever! You know, with that many satellites, Bill Gates is gonna need an *analgesic* with his *Teledesic*!")

They read Parker's comments out loud, and everyone shared a laugh before turning to the sports page.

At the time, the engineers still thought they were headed into the final summer: their last tests before full-up integration of the satellites. Looking down the long scope at a target launch date then set for late October, nothing else in the world mattered so much as the emerging birth of the first two spacecraft, FM-1 and FM-2. Except for their effort to create a consistent pulse, a regular heartbeat that would pass information through the satellite's coiled harnesses, everything else was interference, white noise, bad data, no matter if it was Alan Parker's comic bluffing or Bill Gates's next great enterprise.

Someone tossed the newspaper on a chair when they finished their bowls of chili and tunafish sandwiches. Then they returned to the lab and high bay, no more or less concerned about Teledesic or any other competitor than journeymen carpenters with their shirt pockets weighted down with nails. Without a guarantee that they had jobs for the next year or any real hope of finishing satellite testing before fall, Bill Gates and Teledesic seemed like adversaries of someone else's future, certainly not their own.

A few weeks later another news item created greater anxiety. Orbital announced late in the spring that it would purchase Fairchild Space and Defense Corporation—a historic aerospace company similar in size to Orbital located in Germantown, Maryland—for $70 million in cash and enough shares of stock to make Fairchild ten percent owner of Orbital. It was a startling development that set in motion discouraging changes and unpleasant rumors.

Shortly after the announcement, for instance, Bob Lovell disappeared for a long vacation and a teaching sabbatical at MIT, and soon he was replaced by an equally hard-nosed gentleman from the Defense Advanced Research Projects Agency named Ed Nicastri. Not long after Nicastri arrived, Gurney and Annette heard rumors that the executive staff had compiled a list of people to be laid off.

Those rumors were distressing enough, but they also came wedded with reports that Fairchild's veteran engineers would soon replace or come to manage Orbcomm's young satellite builders. David Thompson, it was said, no longer trusted his own engineers to complete the job on time or to carry on with the production of satellites for the full constellation in 1995. Throughout Orbital the purchase of Fairchild was interpreted as a show of no confidence. Many assumed that Fairchild would take over the entire Space Systems Division and move all the company's satellite teams to Germantown, where satellite production would continue under the auspices of veteran aerospace managers.

Gurney knew enough about Fairchild to think poorly of the purchase. He had worked there for a few months in a contracting job, and like many other engineers at Orbital who knew Fairchild's reputation, he judged the merger to be a "bad fit." Orbital's evolving culture, which stemmed from a busy, youthful, entrepreneurial spirit, seemed at odds with Fairchild's; as one of the first aerospace companies in the United States, it had seen many of its facilities idled by cuts in defense spending, and after a heavy layoff of engineers, it still retained a large caste of middle aged, midlevel managers desperately fishing for projects to manage. Fairchild's old DOD connections, its traditional ties to congressional pork-barrel politics, and its bureaucracy-laden support services for NASA/Goddard made it a proper target for an Orbital takeover. It did not, however, suit many of Orbital's engineers as a property David Thompson should have pursued in their bid for greatness.

"Besides," Gurney told his friends, "there's only one decent place to eat in Germantown, and it's in a trailer park—literally a trailer park—and all they have are sub sandwiches. I went up there a few weeks ago with my girlfriend to help her look for a car, and there were all these Arab guys selling used cars totally covered with new paint—like all the keyholes were painted over. That's all I know about Germantown. Old. Boring. Really depressing."

More changes followed quickly after the Fairchild acquisition. Around the first of June, not long after the announcement, David Thompson called an all-hands meeting with the satellite team—one of the few times he had ever asked to meet with them as a group, so naturally everyone expected the worst. As Rob Denton said, the team went anticipating that Thompson would offer "either a bunch of carrots to make us work harder or a big fat stick across our asses for not working hard enough."

Instead, the CEO took an entirely new approach, a tack so surprising that Gurney could still recall his exact words: "I'd rather have us launch a B-minus satellite in October than an A-plus satellite in December."

The idea, it turned out, was to encourage every engineer to hunt for ways to accelerate schedule by cutting back on the satellite's performance. Ever since then, everybody joked about their "B-minus satellite." A spec was a spec. No subsystem on the spacecraft had margin to spare. The only way to save time was to run through all the performance tests, redline the flaws and failures, and return later to prioritize the ones that had to be fixed, then let everything else slide.

It sounded like a bad choice—to young engineers, a nightmare in the making. But Thompson had seen engineers on other projects cling to their work too many times and he suspected Steffy's team was shooting for perfection. The company needed the next round of Teleglobe financing more than it needed to fly the world's most excellent spacecraft.

Summer should have been a happy time. The team had moved out of its dismal pit at Sullyfield and into the new corporate headquarters at Steeplechase near Dulles airport. They left behind a wreckage of broken bathrooms and scuffed carpets to take up residence in the building that some still referred to as the Taj Mahal.

Even though Gurney managed to acquire an office with a mountain view on the fourth floor, his work grew more difficult as the software integration tests began connecting box to box. He was not at all happy.

Because each engineer had written his or her own piece of code for only one particular subsystem, often without conferring with anyone else, the first comprehensive tests of the satellite's software system turned up hundreds of flaws. Not only did some boxes refuse to talk to others, but during many tests bursts of data from one box would kill the box on the other end. The weirdest, most inexplicable disasters tied them into knots.

As engineers changed their software code to correct discrepancies between subsystems, old versions of code relegated to an archive were replaced by new ones written to repair deficiencies. But as people discovered snafus with new code during testing, they sometimes reinstalled the old code without telling anyone, and suddenly entire rings of boxes died on the spot.

They joked that the latest slogan for the team should have been: "One step backward and two steps . . . backward."

Like Gurney, Annette also began having nightmares late in the spring, about the time of the Fairchild announcement, just as the system-level software tests presented the same continuing, persistent difficulties for her. She, too, slept with the fear of doom, and many days she would come into work, after a poor night's sleep with dreams to tell.

At first Gurney thought Annette's dreams were just hilarious, even though she seemed quite clear about what they signified to her. In her first dream she said that she and Denton had set up a high school science fair using computers and test equipment from the electronics lab and then assembled a large collection of compact discs to assist with the project. But just when children gathered to watch the initial demonstration, the kids started to cry and whine. "We thought you said you knew what you were doing!" they shrieked. "You lied! You lied!"

In the dream Annette turned around in surprise and saw Denton scrounging around on the floor, surrounded by the litter of hundreds of shattered CDs, trying to patch together broken shards with sticky rolls of Scotch tape.

In the next dream Annette saw herself preparing a space shuttle launch. She and Grace Chang and Mark Krebs had stepped aboard the spacecraft hauling a large ring of braided rope. Once inside she noticed the rope connected like an umbilical cord to an Orbcomm satellite on the ground and suddenly realized that their purpose during the launch was to hold on to the rope as the shuttle roared into orbit.

In the most recent dream, Annette found herself at work with all of her original colleagues: Gurney, Denton, Eric, Dan, and Tony. Everything looked normal until she noticed that Denton was with Gurney's girlfriend and Gurney was married to Denton's wife. Then people in the lab began to disappear. Whenever a couple of engineers started working together on the satellite, someone else would vanish, and when Annette went through the building searching for the lost ones, she lost her way, too.

Whenever Annette would come into work and tell these dreams, people laughed at her awful foresight. She became the shaman of the electronics lab.

One day that week Gurney had found her in the lab, as usual, debugging the flight computer code, an endless procedure that for weeks kept her tied directly to Grace's attitude-control electronics, Denton's battery-control regulator, Eric's gateway transceiver, and the flight computer. He thought it was no wonder her dreams usually reflected her tragic link to the whole. But that day Gurney had news of his own.

"I had a dream last night," he said as he sat down next to her and logged on.

"You never dream," Annette said churlishly. She had decided she could be just as coarse and boorish as he,if she wanted to.

"I know, but this Fairchild thing has finally gotten too weird for me. I dreamed they fired everybody at Orbital and sent notices out that said, 'Everybody's gone but you Orbcomm guys.'"

"Wishful thinking," Annette said.

"So I guess we'll get our pink slips in a few weeks," Gurney said.

In fact, layoffs had already begun at Orbital. Eric lost a technician. One of the new hires from TRW in California got a pink slip. Dave Steffy was said to have compiled a list of fresh victims, which he had secretly stashed away somewhere in his office, and some people claimed they had heard him say how surprised they would be when the next-round cuts hit the "core team."

Assuming the worst, Annette had spent the previous weekend revising her résumé and then took off Monday to make phone calls about job prospects elsewhere. She had a bead on a job in Austin, Texas, where her aunt made software for PowerBook computers.

"She only spends two days in the office and gets three days at home," Annette said. "And she's hauling in 55K a year."

"How much experience does she have?" Gurney asked.

"Ten years."

"I could live in Austin," Gurney said.

In fact, he had rummaged through the previous Sunday's *Washington Post* classifieds, scanning for aerospace work. Martin Marietta had an opening in Denver, which he regarded as a possibility.

"So why don't you apply?" Annette said. "It can't hurt to at least brush up your résumé."

But Gurney felt ambivalent. He wasn't sure he wanted to give up his apartment in Georgetown. He had made a good life for himself in Washington. He was still taking his bike out some weekends and shopping for a house. And as much as others might complain, he continued to enjoy some parts of the project, especially the people.

He thought for a moment, then tried to swing the burden back to Annette: "So what's Austin like, anyway?"

Reclining on his sofa, staring at the ceiling of his Georgetown apartment, Gurney wondered what to do with his life. His stomach gurgled. The phone rang. His former girlfriend, Britt, was on the other line, calling from her parents' house in Germany.

"Where have you been!" she asked.

It must have been one A.M. in Germany.

"What do you mean? I've been at work. Where have *you* been?"

Britt had been at a bar. She wanted him to come visit her. Gurney had already calculated that the company owed him at least two months of vacation due to all the overtime, but he couldn't see a way to break away from work just yet. They talked for a while, then he hung up.

Maybe he would call Peter's old girlfriend and see if she wanted to step out for pasta. He knew an all-you-could-eat place in Adams-Morgan.

Instead, he imagined a scheme to make the company buy him a modem that he could install in his living room. That would do well with a new Sun notebook. If he could work at home, then he could be on call if anyone needed him at night or on weekends. The right link would connect him to his network computer at Steeplechase, and then he'd be in business. *Total interoperability*: quite a concept. That way he could spend more time on the weekends working on his Saab, get a line into the World Wide Web, whatever.

The muscles in his jaw pulsed as he thought more about the latest problems with OSX. Lost in thought, Gurney drifted from his vague dissatisfactions and again forgot his problems.

He had no clue just how soon he would come to the end of the line.

As Freshouts left the company, any loss felt like a graduation, with the ceremonial last round of whiskey shots and the final toss of a stinking blue lab jacket into a laundry cart outside the high bay doors. Each departure also left a sense of greater clarity, among those who remained, about the significance of the Fairchild purchase and Orbital's newly structured future.

By September Dan Rittman had left the project to return to the Midwest and work for a cable television company. Mike Dobbles gave notice, boasting that he had taken a 25 percent increase in salary by joining a competitor, another builder of small satellites located just off the Washington Beltway.

Most good-byes triggered a sentimental story. Freshouts recalled the morning Dan finally got his DSPs working—around eight o'clock *Sunday* morning—leaped joyfully into the air, cracked his skull on the lab door, and had to be taken to the hospital. Often they said farewell with unexpected poignancy and a few carefully chosen words. Dobbles in particular welled up with emotion when his time came to say good-bye.

"I learned a lot here," he told me, just a few days before he left. "I came here right out of engineering school, but this is where I became an engineer. It's now pretty clear to me that after Orbcomm, what people think of as the old Orbital will vanish. I'll bet that within six months the whole team will break up, and satellite engineering will eventually come under the direction of a bunch of old guys at Fairchild. David Thompson will continue leading them along a path of mergers with whoever else our lawyers decide can improve quarterly earnings. If there's a ketchup company around that looks like a good deal, they'll buy it. Someday, I predict, the Orbital Sciences Corporation will become Beatrice Foods."

On the Friday before he left, a group of them went out to lunch, and Mark Krebs suddenly refocused the conversation.

"Does everybody remember that Kevin Costner movie *No Way Out*, when he sits down in a bar during the opening scene and you're thinking he's the good guy, until he orders a shot of Stoli vodka? And right then you realize he's a dirty rotten commie traitor?"

Krebs leaned out of his chair to catch the waiter's eye, then smacked Dobbles squarely between the shoulders.

"Get this man a shot of Stoli vodka!"

"It was the perfect touch," Dobbles said later. Krebs was as angry as anyone for Dobbles's desertion, but he had managed to find a way to say good-bye that let Dobbles know he would be missed. "I appreciated that," Dobbles said.

Perhaps no one on the satellite team had come to represent the original hope of the Orbcomm project any more than Krebs. Over two years he had brought a sense of urgency and consequence to the project that even Steffy, trapped by the unreasonable expectations of his bosses, could no longer quite muster. His emotional intensity was infectious, spread like gambling fever to anyone who came to work with him, and struck any of those who might have defied him like a bad case of Hong Kong flu. Since his first day Krebs had ranted in the face of any hapless chump who dawdled during technical meetings and privately hounded engineers whose work proved quixotic or whose attitudes seemed divisive. To the delight of his friends, Krebs stayed in trouble, especially with the moody software ensemble. But he did keep a fire lit under everyone's feet.

No doubt Krebs had come to the project intending to build a career. He made no secret that after Orbcomm he expected a promotion into middle management, perhaps even into the lead of program management. After a few initial doubts, Steffy saw that the quality of his work was convincing. Engineers from the Pegasus rocket program called for his help to investigate rocket anomalies; men like Alan Parker appreciated his salesmanship; savants like Antonio Elias respected his technical skills. Best of all, he had created a respectable attitude-control system out of an inexpensive assortment of parts that many thought would never fly, and he had juiced everything to near-peak performance with a precise set of mathematical calculations and algorithms.

Despite his blustery pretensions, Krebs was an elegant engineer.

"A little bit of a drum-beater" was how Dave Steffy remembered Krebs after he first interviewed, unsuccessfully, for a job in 1989. Over the course of time, however, his drumbeat produced rhythms that many of his younger colleagues appropriated—complex, rebellious, tendentious. Ironically, the drum-beater trait was the one Steffy began to admire most.

But even as Krebs's authority expanded, as it did during the summer of '94 when he took over leadership of the software program in addition to his duties with the attitude-control team, changes in Orbital's corporate structure annoyed him, too. If he had come East to escape a career among the hordes who spent their lives, as he would say, "cranking for Boeing, cranking for Hughes, cranking for Lockheed," he saw the merger with Fairchild much as Dobbles had. The difference was, Krebs's wife was pregnant, their first child was due in November, and now at the beginning of midcareer—no longer the mentored but the mentor—it was time for him to rise, to be acknowledged as a leader, and to win the first series of promotions that would leave him, someday, "cranking" for Orbital.

"Who owns us?" he asked me one day. "We're not a mom 'n' pop corporation anymore. We've just swallowed a whale—a sick whale—and I'm thinking, 'Holy shit! We have a mergers and acquisitions office, and guess what? They do mergers and acquisitions!' You begin to suspect more and more that the job of this company is to build rockets and satellites and—oh yeah, by the way, create a megacorporation."

One fall afternoon, a Sunday at two, Krebs walked into the electronics lab, tapped in his log-on (BLUBBER) and began his fifth hour of work for the day. Across the room Bill Cleary, the new software guru, plodded through his second hour of tests hunting a glitch that kept killing the flight computer. The previous evening they had worked shoulder-to-shoulder on the attitude-control system, which, they discovered, someone had loaded with a segment of code so ill formed that it caused the satellite to fly upside down.

The weekend had almost escaped them both. Krebs, dressed in a T-shirt and green shorts, kept squeezing a green tennis ball between the tips of his powerful fingers and thumb, fattening an already strong set of digits for a rock-climbing venture that had eluded him for weeks.

He stood in front of his computer frozen with a gaze both desperate and solemn, mesmerized by the flicker of a red line the size of a toothpick that blinked and turned ever so slowly above a portrait of the Earth, mimicking the movement of the satellite attempting to find its balance and point its antenna directly down at its terrestrial target. The blinking line might as well have been flicking him a gig, though, giving him the finger. It blinked but otherwise refused to edge anywhere close to its target. Except for an occasional snarl or grunt, Krebs said nothing.

In the back of his mind, he kept thinking about Frank Bellinger, the man who managed the high bay, whose condescending rules about dress codes and conduct around satellite hardware sounded to him like rules a particularly truculent junior high school teacher would apply to a beginning-level shop class. (Strangely enough, it had been rumored that in his early days at Orbital, Bellinger had been a cowboy, like him, who tested explosive devices in an empty lot behind the office, in strict contravention of safety rules.)

That morning when he came to work, one of Frank Bellinger's assistants confronted Krebs, as he worked in the bay alongside the body of satellite FM-2.

"You can't wear shorts in the high bay," Frank Bellinger's assistant said.

Working on a Sunday morning, following a late Saturday night in the lab, Krebs was not in the mood to be so apprised.

"Why not?" Krebs demanded.

Frank Bellinger's assistant told him that Frank prohibited shorts for safety reasons.

"If you're soldering, you could get hurt," the man said.

"As you can see, I am not soldering." Krebs said, "But if you tell me where to sign, I'll give you a release absolving you and Frank and Orbital of any responsibility for my safety."

Krebs turned back to his work.

Frank Bellinger's assistant then folded his arms.

"Doesn't matter," he said. "It's not really a safety issue, it has to do with cleanliness."

"What the fuck—cleanliness!" Krebs shouted.

They argued, then Frank Bellinger's assistant again changed tack.

No, the rule prohibiting shorts was not a matter of safety or cleanliness, he said, it was an even a more critical issue—*discipline.*

Frank, the man said, had a vision of the kind of high bay they would create, not one with lax standards, like the old high bay at Sullyfield, but one that suited the more professional ideals of the Orbital Sciences Corporation as it grew, as it completed the final stages of its Orbcomm, Microlab, Apex, and Seastar satellite projects, as it prepared for production of the next twenty-six Orbcomm satellites and entered the twenty-first century.

"Oh, for God's sake," Krebs crowed, "we're not Toyota! We're building three small satellites, and if we're lucky, we'll finish testing in a couple of weeks and launch before Christmas. And if we're especially lucky"—and this was the blow Krebs knew he could deliver that would really ding Frank Bellinger's assistant—"we'll send all our work out to Germantown and let Fairchild take over production there."

The assistant stomped off.

A little while later, he returned to say that he had talked with Frank and it had been decided that Krebs should be expelled. Bellinger himself would see that Krebs's card code was purged from the system on Monday morning so he would no longer have access to the facility.

"Period," the assistant had said.

Krebs was fuming when he came upstairs and found Cleary still working on the flight computer code.

"I am really pissed," he said.

What actually bothered Krebs was not Bellinger, though, whose threats, he said, only bit with gnatlike intensity, but the prospect that he would, within six months, be absorbed into Germantown, where he imagined entire platoons of Frank Bellingers holed up since the days of Sputnik. And when the time came to redesign Orbcomm and begin the production line, those Bellingers would wriggle into middle-management jobs with Orbcomm, and someone among them would take the one position in management that Krebs had earned, not out of an oversight or meanness or disregard but simply because, as Krebs knew, outside Orbital, Frank Bellingers always ascended in management and saw it as their purpose in life to lord over engineers like him.

He watched the red line blink, pointing away from the Earth but neither moving down toward it nor tilting up at a greater distance.

The problem with real-time simulations, as Grace Chang liked to say, is they occur in real time.

With the meager push and tug of two simulated torque rods against the simulated force of the Earth's magnetic field, the simulated satellite would drift into position as slowly as a cloud on a windless day.

One maneuver could take several hours.

He waited patiently.

"Steffy's wrong," he said at last, not bothering to look up at Cleary, who was digging into the body of the EDU spacecraft lying naked on the table between them. Cleary wanted to disconnect the GPS receiver, which he decided was the culprit that kept killing the flight computer.

"He still thinks this is something you can build in your garage," Krebs said.

He and Cleary had already had the same conversation the night before. Cleary definitely agreed that Orbcomm was more computer than satellite. He could never understand why Steffy hadn't hired more computer jocks than young aerospace junkies and wannabes to build the satellite.

Lately, Steffy had been using the phrase "ship and shoot" a lot, Krebs said, meaning he expected they would send the satellite out to Vandenberg and launch it without conducting a lot of final tests at the launch site. "Ship and shoot" would save them maybe three weeks. The idea made Krebs crazy. Steffy had also told David Schoen one day—or so Krebs had heard—that the satellite team did not need any more engineers. "If I had one more person, I wouldn't know what to do with him," Steffy reportedly said.

Krebs was mystified. Steffy had grown more and more distant from the daily operations. They needed a half-dozen more engineers, at least.

"I see a whole range of problems," he said to Cleary. "It's become legendary that two years ago the guys at Boulder said the subscriber transmitter was done, and then they put it on a shelf for two years until this summer, when Morgan finally tested it and found a whole butt-load of problems. I predict that after launch we'll discover that our worst problems are in the com system."

Cleary said, "Whatever." He kept digging into the satellite.

"It's fucked," Krebs said. "My future is tied to how well these satellites perform."

Cleary was a software engineer, a particularly good one, who worked under month-to-month contracts, like a number of newcomers to the project. He had no real investment in Orbcomm, Germantown, or Orbital.

Posted on the wall, the latest schedule showed that the comprehensive performance tests would end within the week. Then the team would test the integrated satellites—two weeks for thermal vac and thermal cycle—at least one week of which would evaporate as the engineers simply set up their tests. No one had written procedures for the tests yet. Alan Parker still thought they would launch by early November.

When Krebs looked at the schedule, he could see it easily slipping into December, probably into 1995 if anyone bothered to count. No one talked about that, though. Pollyannish schedules meant nothing anymore.

After a while the door swung open, and in walked one of the technicians, recently bribed by beer to work weekends.

"Still having problems with the attitude control?" the tech asked.

"Yeah," Krebs grunted.

"Then you ought to go ahead and marry it."

"Talk to my wife," Krebs snarled. "She thinks I already have,"

The tech brought news from the fourth floor, across the artificial pond that separated the high bay from the corporate headquarters. The antenna would go through redesign while it was being tested in California that week. Tony Robinson had completed deployment tests of the solar arrays, and the hinges checked out exceedingly well. Dave Steffy had sent word that the electrical lab—that "porcine habitat"—needed a thorough cleaning.

"Looks like you should think about what you're doing," the technician said, peering over Krebs's shoulder. "That sim looks like crap."

Krebs moved his mouse and opened a graph, plotting the progress of his simulated satellite's turn. Even the technician saw it would take another three hours to complete.

At that point Krebs announced he was going home to build a fence and tend his garden. At six, he would return and see if the satellite had completed its rotation.

The red line blinked in a horizontal position, 90 degrees off axis. In the little window Krebs had created a quick gauge of the satellite's attitude as it drew data from its sensors, pulsed its torque rods, and decided where it was or where it needed to be.

The message in the window delivered the bad news with only one word: "Indeterminate."

The satellite had no idea where it was.

Krebs murmured something to himself but said nothing to his colleagues as he gathered a set of notes to take home. He would finish by evening, and in the morning the final analysis would begin.

He walked out—clamping down on the green tennis ball, squeezing it tight—as silent as fog.

CHAPTER SEVENTEEN

December 1994 – March 1995

Imperfect Object, Imperfect World

Dave Steffy thought it was a miracle they came as far as they had. From September to October of '94, with a new financial incentive plan kicking in, the team went three full weeks without missing a single deadline—probably some kind of record for the satellite-builders. They put three satellites into environmental tests, running them back and forth to Germantown to stuff them into big thermal and vacuum chambers they called "Bowser" and "the Hot Dog," knocking off tests, redlining problems on their test documents, fixing bugs.

Why wasn't Steffy happy?

One day after a fractious management meeting, he dragged into work with a forlorn look. No tie. Tennis shoes. A gloomy expression burdened his face.

The engineers had seen the same look a few days before, when Steffy wandered into the high bay to watch Tony's antenna-deployment tests. Denton had planned that morning to lecture Steffy about pressing the Boulder crew to test the receiver more thoroughly, but then he just didn't have the heart. Steffy was rubbing his eyes the whole time, and he just stood in the background watching the work. He looked so pained, the only response Denton could call up was sympathy.

Everybody saw it. Even the technicians commented.

Sometime that fall the program manager began to withdraw from the project—psychologically, even physically, it seemed—and turned over the day-to-day operations to his assistant, John Stolte, who called on Kim Kubota and Mark Krebs and Eric Copeland to keep the project running apace. He still attended to technical flaws in the antenna, but mostly he spent his working days running interference against whatever dark forces conspired against them, either in the executive suites of Orbital, at Orb Inc.'s conference table, or lately in the newly acquired Germantown offices.

Steffy reminded himself that he always expected problems in the final stages of the program. On the first Pegasus the team had grown discouraged and complained of being worn to a nub during the last months. Of course, on Pegasus they didn't have the larger corporate decisions weighing them down, they didn't labor under the threat of layoffs, there was no Fairchild acquisition to distract them.

When he had worked at Hughes, engineers would look at Fairchild and say, "What is it they do? Build what?" Fairchild wasn't considered a proper home for satellite-builders—it was looked at as a support system for Goddard, which was what it was even now.

Steffy didn't understand it. His engineers were just as perplexed and skeptical—for good reason.

The only point, Steffy assumed, was that Orbital had to show investors that the company was good for increased earnings. Otherwise the deal made little sense. When he pondered what Orbital got in return, Steffy thought the financials must have been reasonable, at least good enough to impress Wall Street. Of course, that wasn't the deal any of his engineers had signed up for when they came to work—quarterly reports, massive growth, layoffs due to lagging sales. Fairchild was old school. It might bolster Orbital's image with the public, but among his engineers the acquisition looked like a backward turn.

What could he tell them? He had tried his best to save Mike Dobbles. Dan Rittman left feeling bitter and angry. Marriages were rocked pretty badly.

Pictures of Steffy's two children and his wife rested on his desk—new pictures. His baby boy was now three.

He laughed to himself.

Maybe I could go to work for my brother-in-law's cousin at America Online.

He could start rewriting his résumé.

Maybe we won't have to worry about layoffs if things keep going like this.

It was not like Steffy. Cynicism rarely controlled his thoughts. He had always acted rationally, even optimistically. By focusing on technical problems, by going nosedown into a mass of electronic puzzles, packaging dilemmas, weight trades, or the netherworld of RF, he had always pulled the team out of its funk. Technology you could fix; but people put him at a loss. What they really needed, he said to himself, was a better frequency allocation.

Now that was an amusing thought.

He tried to pick himself up with whatever good news they had at hand.

For instance, having heard about the Orbcomm project and its "faster, better, cheaper" approach to satellite engineering, the Office of Management and Budget wanted tours of the high bay. A group of engineers from JPL had dropped in to see the satellite, and expressed admiration. The air force wanted to see what they were doing. NASA engineers always seemed enthusiastic. Vice President Al Gore, one of the very few technically knowledgeable politicians in the country, had mentioned Orbcomm's spinoff, Microlab, in a speech recently, pointing to it as an example of how satellites should be built. Orbital was now on the White House link of space companies to see and high bays to visit, and Steffy often got the call to lead their tours.

He felt good, too, that he had won stock options for Kim Kubota, Mark Krebs, Eric Copeland, and Rob Denton, and that he had created a realistic incentive plan—a masterstroke—just as David Thompson announced no bonuses companywide in July. Financial incentives that he managed to prescribe not only offset the bad news from the CEO but it also revved the project during the fall, putting them on an extremely productive pace. In fact, as he looked across his office, he saw a three-foot stack of freshly completed reports sitting on his conference table. Tied in a blue ribbon, thousands of comprehensive performance tests and quality-assurance checkpoints had been completed and signed, evidence of one strong season of efficient work.

But it was not enough. Never enough.

He had spent six hours the day before in an executive management meeting getting blistered by angry vice presidents who thought he had not done enough. He had come out of the meeting shaken, pale, eyes dilated. And tomorrow he would return to the fifth floor to make another sacrifice of himself, this time for company lawyers who would review his presentation to Teleglobe's experts to make sure he covered the loopholes, every jot and tittle. It would be his duty to tell Teleglobe that a few unmet specifications might degrade performance on the system and a launch delay would push them into 1995. He could do it with two viewgraphs, but with the lawyers on his back, it would take hours to rehearse the performance.

So much could go wrong.

He sat at his desk, motionless for those few moments, thinking: *I'll just put on my dancing shoes.*

But still he did not stir.

When Steffy was a young engineer at Hughes, he once asked his boss how he survived infinite tugs-of-war and internecine battles with other managers in the company. Whenever his boss needed to make even an inconsequential modification in their project, he noticed, the man would enter a series of meetings to seek the approval of other managers—electronics, structures, dynamics, attitude control—until he disappeared in a political morass.

The man never would answer the question directly.

Then one day at a party, Steffy cornered him and asked again: "How do you put up with all that crap? How do you keep from going crazy?"

At that moment, his boss's boss, the division's head honcho, walked in. A gruff little Irishman with a quick temper, the big boss scowled.

"You want to know how we do it?" he barked. "You go home after work and the first thing you do is pour yourself a Scotch."

In the three years since he started his first satellite team, Steffy never fell into the habit of drinking. He relied instead on a sense of humor to get him through. Occasionally he would try his own hand at engineering—go head down into a technical problem—to hold him steady, but usually he tempered his moods with whimsy.

Neither caustic nor condescending, Steffy's humor amounted to everyday observations about the fundamental absurdity of engineering. On the worst days, one-liners popped out by the dozens:

"A holiday is one of those days when the mail's not there when you get home from work."

"The engineers on Orbcomm like to say we work half days. If you come to work at eight and leave by seven, you count yourself lucky because you're getting out an hour early."

"You've gotta wonder why we're so eager to build these things. Once people start buying little handheld communicators, nobody in the world will be able to take a real vacation anymore."

"The only thing worse than dealing with an aerospace vendor is working with a construction contractor. First you pay them an expedite fee and then add the 20 percent lie to that. You can set your watch by it."

"You can always pick out the contracts guy during a design review. You're sitting in on this technical discussion about the lubrication of a gear train and some mechanical guy is up there hard out answering detailed questions about his nice, tight technical design, and suddenly some guy walks in and sits down in the back of the room. He's wearing a teal-colored sports coat with a brown shirt and black tie, a pair of checkered sports pants, and $300 loafers. You see everybody glancing back to get a look, and you think, 'Yep, that's the guy they warned me about.'"

He also told stories: about the rocket somebody accidentally loaded with test-mode software—at launch, the vehicle rose a hundred feet and then made a hard right turn and started flying up the beach; about Dick Parfit, the legendary engineer at Hughes whose only job was to draw wiring diagrams of satellites by hand, an art so prized and indispensable, before the advent of personal computers, that it was said the CEO of Hughes would drive Parfit home every afternoon, mix him a drink, and get his slippers. Even after Parfit retired, wiring diagrams at Hughes were still referred to as "parfits."

In one way or another, Steffy understood the awful ordinariness of aerospace, the profession's *Far Side* illuminated by absurdity and weary indifference.

One day when the team met for lunch, Steffy recalled the satellite-control center at Hughes where, as an intern, he had expected to discover the sophisticated hub where a few men controlled the ever-drifting orbits of a dozen of the world's most advanced communications satellites.

"A room not much bigger than my office," he said. "You go in there thinking you'll meet these gurus on the consoles, but instead you go in and here are a couple of guys who look like plumbers—*retired* plumbers—and they're sitting around, tuning in to the Playboy Channel."

One day a college physics professor came to visit the company pursuing an interest in the dynamics of decaying orbits—how satellites drift out of position—and asked to see the control room. He arrived expecting to tease out secret algorithms the company used to offset the decay.

"So the professor comes in thinking our engineers have looked at all the angles and prescribed the most efficient method for commanding a thruster at the dip of a figure eight when they're sinking in their orbit. He asks all these questions about what he thought must surely be a very precise maneuver, and he waits to see how we've scheduled the exact moment to fire the thrusters—you know, to preserve the most amount of gas and take advantage of solar winds and everything else.

"So he stands around for a while and finally asks straight out how the calculations are made and how the timing is deduced and how this was actually done. Well, the guys look at each other, and one of them says there's only one real rule of thumb: 'We fire thrusters about every two weeks, but we just don't do it on weekends because nobody wants to pay us overtime.'"

And that's the way it is in aerospace, Steffy would say.

Someday, when you meet your first outside customer, and the customer says he doesn't give a shit about your satellites or your software or your launch capabilities, all he wants is revenue, the whole business takes on a new perspective. A satellite to an engineer is a wondrous object, an art form based on orbital mechanics, elegant algorithms, and electromagnetic magic, a hand-crafted, thermal-coated gem. But to everybody else in the world, it's just a really high antenna. The fact that you can't even see the damn thing is just fine. To most eyes, it's ugly as hell, and as long as it keeps taking pictures of hurricanes, sending clear pictures for the TV, and bouncing signals to everyone who wears a pager, who cares?

As long as it works.

So Steffy, the incorrigible optimist, could always talk himself and most of his engineers out of any black mood or sense of doom they felt at the end of a day. To him, there was always a bright side, and tomorrow would always prove infinitely better.

But for the first time, during the final weeks of 1994, just three months from the final countdown, it was increasingly apparent that Steffy had run out of jokes and one-liners. Day by day the project leadership fell, in large part, to Kim Kubota, Mark Krebs, and the few remaining Freshouts who had started with him in 1992.

As Steffy continued to withdraw, humors turned black.

Four days before Christmas someone snuck out at midafternoon and slipped back into the high bay lugging a cooler loaded with icy beers. Despite their frantic pace, the team could see an end to a testing program that resulted in more mangled code, fried microcontrollers, and broken hardware than anyone ever imagined.

With a late evening ahead, Mark Krebs was thinking more about the team's morale than about its efficiency, for that was how he had begun to think in his role. Increasingly, he was a team builder, an ex officio manager for team spirit and cohesion. A few free beers, he assumed, would bring everyone together for an hour or two.

Strolling through the high bay, Krebs occasionally stopped at various test sites on the floor to consider the progress. He still had to conduct final magnetics tests in a far-off corner at the back of the bay, but at each site he paused to whisper to his teammates: "Cold brew in the conference room at eight."

Rob Denton and Eric Copeland were commiserating with Nancy Cohen, their new contractor from Booz-Allen, about the breakup with her boyfriend. Nancy watched RF signals dance across the screen of a spectrum analyzer. Dressed in faded blue lab smocks, they were shoeless and bleary-eyed. The three of them looked more like nineteenth-century Chinese railroad workers than twenty-first-century satellite engineers. They sounded miserable.

"Are you ever happy?" Denton asked.

"I'm never happy," Eric said.

"I want to go home," Nancy said.

Upstairs in the conference room, Tony Robinson had John Stolte, the team's assistant manager, up against the wall, complaining about his inability to assign a design engineer to resize the antenna trough.

"I'm pissed!" Tony said. "I get tired of beating my head against the same wall. How many times have you heard me begging for a designer?"

"A few," Stolte said amicably. Stolte was a very nice guy.

"A few? I literally have to steal one about every six weeks from some other project. I've got to go begging to get someone for a couple of hours of work! I've had it!"

"I'll find someone for you, Tony," Stolte said.

"You know—besides, I'm pissed because—it's not just the lack of good designers," Tony said.

"I know. Everybody's upset about Fairchild."

"Let me show you something."

Tony reached into his briefcase and pulled out a sheaf of papers, the latest organizational chart from Fairchild, which he had lifted surreptitiously on a recent trip to Germantown.

Tony sniffed, pointed at the Orbital logo, which had replaced Fairchild's flying horse emblem, then opened the pages.

"What do you see here?" he said.

Stolte wasn't looking.

"Line after line of managers," Tony said. "C'mon, look at this! More managers than workers, right? They're not making any changes in Germantown. They're waiting for us to come over and fill the gaps under all these geezers."

"No one knows what's going to happen, Tony. You can't anticipate anything yet."

Tony tossed the papers back into the briefcase.

"Most of us came to Orbital because we wanted out of the big companies, right? You did, too, John. And you can say whatever you want about the schedule, but at least until now we've had the freedom we were guaranteed to be creative and manage the project ourselves."

For Tony, it had always seemed like David Thompson's vision for Orbcomm had worked.

"I agree, it's been working lately," Stolte nodded. "It's really starting to come together now."

"But look at their org chart!" Tony argued. "Over there you're just another little cog, you're expendable. Either that or you find a way to squeeze into a little midmanagement slot—after the old guys retire—and then you spend the next twenty years hiding out under government contracts."

"It's not that bad, Tony," John said.

"The last piece of work they bid? I heard they spent six months just drawing up the contract."

Everyone thought they knew about Germantown. After weeks there in the fall conducting electromagnetic emissions tests and thermal shock tests, the image of the former Fairchild plant as a retirement home for Apollo-era bureaucrats was verified in the minds of many of them. One afternoon they watched in shocked amusement as an older female employee sporting a beehive hairdo walked out early carrying a bowling trophy. They saw workers at Germantown playing card games on their computers during the day. On one occasion, despite the satellite engineers' need to use an empty test chamber one afternoon, the Fairchild manager in charge of the testing area refused to open it because he had scheduled the chamber for a government customer—"who pays the salaries around here," the man reminded them—*two days later.*

They endured so many thoughtless restrictions and heard enough orders barked at them by retired military officers acting as Germantown bureaucrats—gray-haired, slack-bellied men wearing drip-dry polyester shirts and pants, speaking in southern accents—that they raged endlessly afterward.

"A telephone rings around there, and everybody looks surprised."

"The lights go out automatically at exactly 5:18 every afternoon, and suddenly you realize the entire building's deserted except for whichever one of us is still around testing Orbcomm shit."

"The place looks like a minimum security prison."

"Spend just one day at Germantown, and you'll feel young and thin again."

John Stolte was not unsympathetic. He agreed to find Tony a designer and, in addition, offered to bring a few of Germantown's best technicians over to Steeplechase to help him out for the final weeks of mechanical testing.

Victorious, Tony went back out into the high bay to work on the satellite antenna, a 100-inch-long cylinder made of graphite webbing covered with Kapton tape that people affectionately nicknamed "the Kapton condom," for obvious reasons.

Before the end of the year, Tony figured he would have ten or twelve engineers working just for him.

By eight o'clock Krebs stood frozen with a group of other blue-jacketed engineers, Grace Chang, Paul Stach, and Willie Lieu, watching a strip chart slowly unfurl at their feet from the mouth of a variometer, a machine monitoring the direction and strength of magnetic dipoles on the satellite. Orbcomm satellite FM-2 sat perched atop a pine table 20 feet away in a far corner of the high bay. Cordoned off with a yellow plastic chain, one entire quadrant of the high bay had been claimed for the final magnetics tests, which would all but conclude a comprehensive analysis of hardware in the satellite's attitude-control system.

Krebs ripped off a section of a strip chart and marked it in ink, drawing arrows to meandering trends in the chart that looked, to him, like an indication of mistakes in the test setup. His notations reflected a lighthearted mood—*"Here's where the subscriber transmitter starts smoking. . . . This is humorous! . . . This thing's not broken, it's uncalibrated!"*

Finally, he concluded that something was wrong with the test itself, and directed his teammates to tune the procedures. They could take their time, though. He expected them to be there most of the night, regardless.

Weeks later the conference room was still littered with beer bottles, the walls pocked with tiny black nicks from flying caps. A coffeepot in a corner held a tarry sludge in its bottom, accommodating the growth of a furry white fungus, and on the larger markerboard that displayed the most up-to-date version of the schedule, someone had erased the first two weeks in March and replaced them with a big question mark.

The new boss kept asking Krebs questions like "Did you degauss that mother-sucker?" and "Won't the problem go away if you degauss that mother-sucker?" and "Are you sure there's no weird magnetization on the power supply rack that's gaussing that mother-sucker?" No matter how many times he asked, no brisk interrogation could motivate Krebs any more than the pure terror he already felt since, after one simple oversight, the attitude-control team had steered the team onto the path toward disaster.

"If we don't launch by mid-March, we don't launch at all," Steffy told them all during the first week of 1995, echoing the stern warning David Thompson had given him over Christmas. "Orbcomm will die."

The cheeriness of the beer party in late December seemed like a forgotten promise by the start of the new year. No one would have imagined that the satellite batteries could turn into permanent magnets, least of all Mark Krebs, whose colleagues Grace Chang and Paul Stach had tested the satellite almost a year before and concluded that the magnetic properties inside the spacecraft were benign, slight enough to have no effect on the spacecraft's stabilization on orbit. But during the previous year the batteries had not yet undergone hundreds of cycling charges and discharges in tests that, in the meantime, apparently created a perplexing change in their internal chemistry and magnetized the batteries.

Few people on the team even remembered the first magnetics tests, which Grace had conducted one cold autumn day in the parking lot outside. The problems they identified then, and simply named "weirds" during Christmas week, now seemed to have been waiting to ambush them for months.

Alan Parker and his alarmed honchos at Orb Inc. naturally expressed concern—their usual paroxysms and open disdain for Steffy's engineers—and started calling the satellite team a "day care center for amateur engineers." They said anyone should have known that nickel in a nickel hydrogen battery would magnetize if you sent repeated surges of electric currents through it. Any fool would have known that.

But as Mark and his teammates gathered in the big boss's office in early January, joining Ed Nicastri in unproductive conversations about their "mother-sucker" batteries and holding phone conferences with battery experts around the country, the question as to why those six little nickel-hydrogen numbers magnetized never became clear. Some batteries magnetized, others did not; some cells showed a dipole lined up along one axis, others showed them lined up along another. It defied reason.

Even the scientists and engineers at Eagle-Picher who had designed and built the batteries had no explanation. "They probably magnetize all the time on other satellites," one expert concluded, "but because other satellites are so much bigger, nobody noticed."

The one unalterable fact of life: No one else in aerospace cared. Most satellites weighed so much more than Orbcomm and contained such powerful control systems and orbited at such greater distances from the Earth that a magnetic charge had no effect. Only on a lightweight satellite like Orbcomm, which scaled in at less than 95 pounds, orbited well within the Earth's magnetic fields, and relied on a passive control system comprising little more than a couple of bar magnets and thrusters capable of firing only the smallest of mouse farts, could the torque of dipoles with all the combined strength of a simple refrigerator magnet send the satellite veering off balance. That was it. The slightest nudge against the Earth's magnetic field, the gentlest tug, and the spacecraft would wobble out of control.

Naturally, no one took the news harder than Krebs. Over Christmas while he continued testing in the high bay and analyzing the improbable results that kept curling up around his feet from hundreds of strip charts, he hosted unannounced visits from company executives, who dropped by and spent many days letting him know that the problem required the most intense scrutiny. As if he didn't know.

Forced to endure continual oversight from bosses who apparently sniffed the rank of failure from across the gully separating the high bay and executive offices, Krebs felt pressures he never experienced before. While Nicastri leaned over his shoulder nervously jangling pocket change and car keys, Krebs's stomach churned.

By January he had salamander eyes—puffy, red, moist—from so many sleepless nights. A head cold choked his voice into a smoky wheeze. Knowing little about either magnetics or batteries, he ramped up quickly by studying a variety of textbooks on the subject, then appointed two separate teams of engineers to assault the problem at different angles—a "science fair" team that would conduct experiments on the batteries to identify the cause of the magnetizing (gaussing) effect, and an engineering team that would pinpoint and analyze magnetic forces cropping up almost randomly inside the satellite.

As they chatted in Nicastri's office one morning, the new boss finally told Krebs, "The checkbook's open."

Immediately, Krebs sought the best consultants he could find in the United States. Unfortunately, the gurus he located knew little about magnetic effects in batteries.

He talked at length to a battery expert from TRW one day, questioned a group of engineers at NASA another, held conference calls with the most prominent battery experts in aerospace. But the consistent refrain was that Orbital should simply ditch nickel-hydrogen batteries and replace them with batteries that would not magnetize—silver-cadmium, for example, or silver zinc.

The problem was that any new battery would have to be built to specifications. Silver-cadmium batteries, for instance, developed for an experimental program at NASA, did not exist in the right proportions to fit Orbcomm's spacecraft shelf. Silver zinc was available but would only serve a lifetime of one year in orbit. Most daunting of all, a change in batteries required a wholesale revision of software across the entire spacecraft. Essentially, the best answers required a six-month development program, a solution so absurd as to merit not more than a moment of consideration.

One day in mid-January, out of desperation, Krebs turned to the Internet. He discovered an electronic bulletin board for an aerospace propulsion group, typed out a statement of his predicament, and posted the note seeking help. Dozens of people responded. Most of them seemed more interested in UFOs and aliens than in satellite attitude control, but a few resulted in productive leads.

Finally, one response sounded promising: a small, secretive aerospace company that manufactured materials for the U.S. military ("some spook outfit," Mark said) produced thin sheets of a highly permeable soft metal that glistened and crackled like aluminum foil, a material that would wrap around the batteries and potentially shield the dipoles so their magnetic forces would be ineffective.

"Classic black magic," Mark called the stuff, "with a supreme affinity for controlling flux leakage from almost any magnetic source."

He managed to acquire a small batch of the metal from someone inside NASA, a limited roll measuring only two meters square. Needing a larger supply, he searched the roll for any clue of its manufacturer or its composition and found instead that its identifying tag had been stripped off. Thus, he named the obscure material for the man from the "spook outfit" from which it came, a mysterious Hispanic fellow, who had assured Mark that he could provide more, if necessary, for a price. They called it "Mario Metal."

The engineers wrapped batteries with Mario Metal the entire third week of January, swaddling one black battery in silver foil, then another, set them up on the pine platform, and swiveled each battery individually 360 degrees as they took measurements. To reduce the likelihood of interference from cars passing through the parking lot or airplanes passing in and out of Dulles airport or any other electrical noise that would skew the data, they worked mostly at night. The optimal working hours fell between midnight and four A.M.

Unfortunately, just as the three completed spacecraft were about to undergo their final integration tests, Krebs stripped all the batteries out of both Flight spacecraft and the Qual.

Learning that he would have to conduct satellite integration tests without the batteries, one of the test engineers, Kirk Stromberg, went directly to Steffy and begged for more time. It was imperative to redo the integrated tests once the magnetics problem was resolved and the batteries went back into the spacecraft.

Kirk cornered Steffy in his office and pleaded for more comprehensive tests.

"If you want to know why I'm annoyed," Kirk snarled, going right into Steffy's face, "it's because we often find ways to rationalize shortcuts in the test schedule, but we always find out later that something has gone wrong."

"So Kirk, you think you're getting anal about testing full-up configuration?" Steffy asked calmly. "What about the final software load? Couple of missed keystrokes, someone loads a file out of the wrong directory, we don't have a spacecraft. I can build a rationale that we'd better spend our extra days there—if we had extra days, which we don't."

Kirk wouldn't back down: "But I can show you that every single time we've added one more functional test to the satellite, we've found a bug or something else we needed to learn."

"Can't do it," Steffy said.

"Why not?"

"Because someday real soon, somebody around here's going to really pick a launch date—probably in mid-March—and whenever we lose a day to an extra test, our final fourteen days to launch will really be thirteen, and then we're in trouble. Believe me, there are other pressures driving us now."

Kirk nodded and mumbled that he understood, then left. But he, like many of the other engineers, did not trust Steffy completely anymore.

Increasingly, the engineers suspected the program manager was no longer willing to fight with executives on the fifth floor. They noticed that he rarely bothered to explain the "other pressures" he referred to so often, and they weren't willing to buy his line that he simply wanted to protect the team from the nervous agitation at Orb Inc. and the sometimes-harsh criticisms of Orbital's chief executives. His engineers concluded, for the most part, that Steffy was gutless.

He knew what they thought.

If the wall on one end was made of tests and technical problems, the wall at the other end was Pegasus. With the magnetics problems drawing every reserve, the game was over.

"I just don't have any cards left to play," Steffy told his assistant John Stolte that same afternoon, explaining the confrontation with Kirk. "With these magnetics problems you don't know when we'll get done. You're out there every night trying this and that. We have to go through who knows how many more days before we find a solution. And once we find it, how many days will it take to finally wrap the batteries and rewrite the software? Two or three are fine, you can do that by putting the team on three shifts. But when it's two weeks and counting, ten days and counting, we've got no more margin. I'm afraid there's no way out now. There's just no way."

In fact, he had realized sometime in early January that they had no time left to broker. But he had kept the news from everyone—he still avoided the lab and high bay—and spent his days charting the team's progress from a distance, scanning the schedule to find some way that he could reshuffle their deck of remaining days and sort out a more efficient plan of attack.

He had already managed to push the satellite's launch date as far down the flight manifest as possible so they could bump the schedule into 1995. But they could not push past March. Pegasus launches guaranteed to several U.S. government projects scheduled for the spring would amount to real revenues for Orbital, unlike the Orbcomm project, which would grow slowly over a period of several years. Delaying launches with the government would affect future sales—a situation deemed untenable by the executives. Looking further down the ladder of the company's launch dates through the year, Steffy saw no other openings until late summer at best. At that point the question was, very simply, who would financially support the project? The realistic answer was, no one.

A host of Orbital's executives assurred him that neither Teleglobe nor Wall Street could tolerate another delay. Orbital's stock price, deflated below 20 points on the NASDAQ because of the many rescheduled launches, would take a devastating tumble. Most analysts in the aerospace business regarded Orbcomm as the single project that would boost Orbital Sciences into the big leagues. The company's future rested on the satellite-builders. Teleglobe, though interested, had not fully committed and could back out easily if the satellites weren't in orbit soon.

And so it was settled—Orbcomm would fly in March.

Everyone was frustrated, the satellite engineers still floundered late in the test flow, and Steffy had nothing left to juggle.

"We're at the point where every engineer wants to do something for the last time," Steffy told me one afternoon. "They want to do the last thermal cycle, the last screw tightening, the last connector mate, the last blanket wrap. . . . Everyone wants the next procedure to be really, really, really the final one. But we're just not there yet."

It was this final loss of time that disturbed Annette Mirantes's sleep with hairy nightmares, that gave Mark Krebs salamander eyes, that forced Dave Steffy into isolation, and that resulted in David Thompson taking an implacable stand on almost every request.

The lack of margins and the subsequent risks finally would yield, as Grace Chang noted one day, "one more imperfect object for an imperfect world"—as clear a statement of their anguish and dilemmas as anyone could make.

As testing continued, they had learned to expect the worst, no matter how close they drew to the final days—just as any team of veterans might. Out in the high bay, within a circumference of 30 feet, three Orbcomm satellite bodies lay open, surrounded by tall, red modular racks stuffed with test equipment and computer monitors. The raspy notes of Bruce Springsteen's voice and his stinging guitar echoed from a boom box across the gigantic white hall while engineers conducted their final tests in Virginia.

"Shit, this is terrible!" Rob Denton said as he read a spate of error messages scrolling up at lightning speeds from the Qual satellite. "We're losing three out of every ten bits."

"Three out of ten is acceptable," answered Eric Copeland, who had paired up with Denton for final tests of the gateway transceiver. "We're only building a B-minus spacecraft, remember?"

"Oh, yeah, B minus."

"Oh yeah, B minus," Eric repeated, mockingly.

A few feet away, perched atop a stool behind his own PC, Krebs cracked his knuckles and griped. He had intended to use FM-1 in several tests of his attitude-control system that day. But the full-up, integrated FM-1 satellite lay nearby, bolted shut, neatly tied down atop a table, while FM-2 lay open before him, eviscerated by engineers who had stripped out several primary components for tests in the electronics lab upstairs.

"One full day of downtime," Mark muttered as he fitfully popped open screens on his PC searching for something to do.

Out on the high bay floor, with the addition of engineers from the Fairchild acquisition and a stream of contractors who had come in temporarily to help with the final stages of testing, the satellite team had grown to more than seventy-five people.

A pair of eyes followed their progress from the glassy enclosure of the second-floor conference room. In her skybox overlooking the massive bay Kim Kubota—resident circus master of the final satellite-build plan—watched. Her dark gaze was framed beneath a brow weighted with worry and concern.

For a few moments she watched as an engineer on the floor below her slipped an X-acto knife along the outside rim of FM-1, slicing neatly through a thin ribbon of silver tape that skirted the edge of the solar panel. The engineer wedged a small screwdriver beneath the panel at one of the three brackets and attempted to pry the panel loose. The panel held tight.

Several more members of the team gathered around, like a highway crew at rest, apparently waiting to see who would be the first to spill blood on the satellite. Finally, someone reached for a pocket knife and carefully slid it beneath the panel, angled the blade left, then right, and the top popped up.

She saw one of them motion to a nearby technician, who immediately came over, unscrewed the satellite transmitter from the spacecraft, clipped its wiry connections, tenderly lifted it out of the body, like a surgeon removing a kidney, and gently set it on the table.

The natural progression of tests, which should have set a rhythm to testing and analysis that led logically from EDU to Qual to Flight Models 1 and 2, never had a chance, Kim thought. Vendors neglected to ship components on time. Parts failed during early screenings. In the "shake 'n' bake" of environmental testing, electronics broke. Pieces of the EDU satellite or Qual that failed had to be refabbed and, at times, redesigned. Now, with only a few weeks left, the engineers found themselves in the absurd position of conducting risky tests on the actual flight satellites before seeing how such tests would affect the trial vehicles (the Qual and EDU). Instead of completing tests on the EDU before moving on to the Qual and finishing with Qual before building FM-1 and -2, the engineers often tested the four satellites in parallel, forcing themselves to accelerate trials on the flight vehicles before completing the same tests on the Qual or even the EDU.

To confuse matters even more, members of the Microlab team, whose Lightning Sat and meteorological instruments would launch with FM-1 and -2, had started stealing whatever pieces they needed out of the Orbcomm satellites to integrate into their own. Seeing their four satellite models robbed like Egyptian tombs during the night, Orbcomm engineers were left to test their systems sometimes with an amalgam of spare parts. At times their test equipment disappeared, too, appropriated by the Microlab crew, whose deadline posed limits just as severe as their own. Pilfering equipment had become an accepted practice, though a particularly desperate one, that further exasperated the senior staff.

Somehow, on most days, Kim managed to keep peace. But today, as she watched the unexpected surgery taking place on the floor below and then met with one person after another who came in to complain that the problem would require days to resolve, she exploded.

"I'm running a refugee camp here!" she exclaimed. "I am Zaire!"

Since the previous summer, her hours at work had straddled two twelve-hour shifts ranging from six in the morning to well after midnight the next day. She convinced her friend from the TOS mission, Kirk Stromberg, to join her, and together their accumulation of hours in the high bay and its conference rooms staggered even the most devoted Freshouts.

After several months on the job, Kim and Kirk formed a social group for the team they called "the Losers Club," an assortment of unmarried Freshouts who found themselves now so enslaved by their jobs that they no longer had time for friends or socializing in the outside world. Whenever the weekend rolled around and they managed to find a break for a few hours, members of the Losers Club organized themselves for a meal and, on rare occasions, a movie. Mark Krebs, although married, was made an honorary member of the club, they said, because he seemed like such a loser anyway. The club was a well-kept secret from engineers on other projects, though, and managed to form a bond between its members, who in a peculiar way seemed to begin enjoying their sacrifices and ritualistic jaunts on the weekends.

The fact that Steffy knew nothing about the Losers Club told the full story of where they had finally come as a team by 1995. Kim had established herself as an empathetic leader, standing in more often at staff meetings, holding together an increasingly exhausted and hostile team. If she felt like Zaire, offering sanctuary for young engineer-refugees, it was no wonder.

When she finally decided to go to the floor and investigate, Grace Chang was peering down at the satellite transmitter, spread in pieces across the table.

The first thing she noticed were the three white fly wires coursing the top. Then she noticed that the board was hot. During tests, someone said, it looked like the transmitter actually was trying to receive.

"How in the world did this thing pass acceptance testing?" Kim demanded. She looked at the faces of engineers and technicians grouped around the satellite.

"Good question," one of the technicians said.

"How did this thing even get past component-level testing?" she asked again.

Some of the engineers, seeing a serious oversight, backed away, moved to cluster by the test equipment, and whispered among themselves:

"I think they just forgot to apply the last fly wire," one said.

"Right. So?"

"Well, I heard that Boulder knew and still signed off on it."

"Oh, that's bad."

"What?"

"You mean, it was like, 'Where's the fly wire? Oh, yeah, one of them got left out. Oops.'"

"Yeah, B-minus."

Kim suspected that someone in Boulder's lab must have managed to run the transmitter through acceptance tests without kicking it back to be fixed. The politics of that—well, she didn't even want to contemplate it.

Now she would have to decide whether they could dismantle the transmitter for repairs without sending the entire box back through a time-consuming round of thermal cycle and vibration tests, or do a thorough job of repairs and retesting, which might cause them to lose another two weeks and bump launch.

"All we need is a tiny fly wire," Grace begged. "One fly wire. The tiniest of tiny fly wires."

Kim could have called Steffy and forced him to sign off on a quick fix. Even one lost week, she could argue, would spell disaster for the schedule.

She collared the technician and again questioned him over the open box. "Didn't anyone know what the test plan was for this thing?"

The technician shrugged, Kim slapped her clipboard shut and ordered them to dismantle the box. The technicians could make the repair overnight, and she would find someone to send it out to Germantown to move it into thermal cycles before the next morning.

"Fill out a noncompliance report," she told the technician, "and get me a work order for tomorrow."

She turned to Grace: "I'll call Steffy to get his signature, and in the meantime I want you to get the schematics sent to Boulder—immediately . . ."

Then she turned to the technician, glared, and demanded an answer. "I still don't understand how that box passed acceptance testing," she said.

But the technician said nothing. If he knew anything at all, he wouldn't betray a confidence. Sometimes silence was the only honorable response.

Kim trotted out of the high bay in a huff, back to the conference room to review her schedule charts. Mark Krebs, who had never once commented during the entire confrontation, sat at his terminal laughing.

"Another night at the satellite bar!" he crowed. "It's late! We're drunk! We lower our standards!

One week before launch, at lunchtime one day, Mark Krebs walked into the cafeteria, grabbed a bottle of cranberry juice, and sat down at the table across from Dave Grossman, the young engineer who had written the flight software for the Apex satellite. During the summer, Orbital had launched Apex, its first satellite, and immediately the spacecraft had hung up due to a software glitch.

"Give it to me briefly, Grossman," Krebs said. "What went wrong?"

Grossman actually looked reasonably happy since he was the one who had identified the anomaly from the ground and then quickly settled on a way to fix it. A quick software upload had saved the mission.

"I never got too excited about it because after the launch I was just waiting for something to screw up," Grossman said.

The problem turned out to be simple—the sun sensor spewed bad data.

Krebs wanted more detail.

Okay, Grossman said, what happened was, whenever the attitude-control system saw two bad packets in a row, it immediately reverted to a contingency, safe-hold mode. Unfortunately, with a bug in the code, the safe-hold mode sent a ripple through the whole system causing the solar panels to keep trying to deploy even though they already had deployed. Then the battery regulator reacted to that, and the whole satellite went berserk.

"The funny thing," Grossman said, "is the company that built the screwy sun sensor was still trying to fix hardware problems the week before we launched."

Krebs hung on to every word. He leaned over the table.

"What can we do to avoid these things?" he whispered.

"What do you mean?" Grossman acted surprised. "You know, Krebs, you can't think of everything."

"I mean, what can we do right now, as long as we're still testing our systems on the ground, to keep anything like this from happening to us."

Grossman looked even more amused. "You can't do anything," he said.

But Krebs only became more agitated.

"This is my nightmare, Grossman," he said. "Can't we write something into our code that will protect us from the kind of thing that happened to you? I mean, isn't there something you learned that we could benefit from now?"

Grossman, naturally, smiled this time. People rarely saw Krebs express doubt.

He thought a moment, then shook his head.

Krebs picked up a napkin from his lunch tray, ripped it into three sections, and laid them out on the table.

"Here," he said, leaning closer, "one, two, three. We've got all these different teams building software, and the problem is, it's never been completely integrated. Everybody's got a different way of building it up, and no one's spent time thinking about how it all fits. I mean, are we entering a nightmare phase here or a long nightmare phase?"

"We never had a systems engineer on Apex, either," Grossman said. "There was never any plan."

Nothing he said could satisfy Krebs. It all sounded too familiar. Krebs nodded, unconvinced, then said he wanted to hire Grossman into the Orbcomm program.

"You can start next week," Krebs said, as if the deal were done.

"I'm working for Seastar!" Grossman declared, completely taken aback.

"Seastar?" Krebs looked puzzled.

"They're struggling," Grossman said.

"So does that mean we're next in the box?" Krebs asks.

This time Grossman finally laughed out loud.

"You've always been next in the box, Krebs. Didn't you know? You guys are the next ones to launch."

Silently, absently, Krebs twisted the pieces of his napkin into little bits and scowled into space, an expression so complete and sincere, it was almost disturbing to witness. Then he stuffed the shredded napkin into his juice bottle and rose to leave.

The conversation had been so quiet that no one else had heard, and so undemonstrative that no one sitting around them seemed to have noticed.

Krebs picked up his tray and walked out of the cafeteria, back to his magnetics nightmare, the team's last great challenge.

The Losers Club met for the final time on the East Coast, Friday night, March 3, 1995, during the satellite's last tests in the high bay. At three A.M., Kim Kubota took a break for a soda and fell asleep outside the high bay doors.

The team had complained all day because, they said, Steffy hadn't sprung for dinner in months, and because he left for home early to do his laundry. They complained about the Microlab team because its engineers sent out for pizza at suppertime and never offered any to the Orbcomm team. They complained that David Schoen's engineers from Orbcomm Inc.—now referred to, disparagingly, only as "OINC"—had taken over the satellite team's lab, and it quickly became clear that Schoen's people were nowhere near ready for a launch.

About three-thirty A.M., Kim stirred awake and noticed the rest of the club gathered at the foot of the soft-drink machine, apparently waiting for her to decide if they should simply skip the last few tests and go home for a decent night's sleep or tough it out with her and keep working until the job was done.

Kim looked at them, folded her hands, and said, "Dear God, don't delay this launch."

"So we can skip the last tests?" one of the Freshouts asked.

"Ship it," Kim said.

"Command decision," the Freshout announced to the rest of the team. "Kim says we can pack it up and ship. Think B minus, everybody."

"No, think D," Kim said.

They argued with her—the satellites weren't that bad.

"Don't worry about it," she said sardonically. "D is passing. You get a D in school, you pass. You'll go on to the next class."

Yeah, but the satellites are at least a B minus, they said.

"No, D," she said. "You pass. You will advance to the next project."

She lay down on the floor again, was quiet for a minute, then got up again.

She stretched and picked herself up. They were still looking at her for the next order.

"Okay," she said. "You're right. Forget D."

The truck would come for the satellites at noon. With eight full hours before the spacecraft left for California, they still had time for a few more tests.

Bulletin Board

"FCC Commissioner Rachelle Chong, drew laughter from attendees when she followed the remarks of the other four commissioners by using a Valley Girl accent to call the Orbcomm agenda "totally rad." She followed that by adding, "It's about time we had a cool satellite technology."

Satellite News
October 24, 1994

"I'm going to business school in the fall. Eric and I have applied to the same schools so, you never know. We've really talked a lot about starting our own company. Look at these vendors we hire. I mean, these guys make millions of dollars building com equipment for companies like this. Dave Steffy's even asked us if we'd like to go into a start-up with him. I'm not kidding. There just aren't that many really good ones out there. Believe me, we know."

Rob Denton
Satellite engineer
October 1994

"Investment manager Joan Lappin has latched onto a stock that's in orbit. She purchased a load of shares of Orbital Sciences, which got government approval on Oct. 24 to launch a satellite for worldwide paging services. That means, says Lappin, that Orbital can now jump-start a satellite system that will offer cheap two-way paging and messaging services at any point on the globe."

Business Week
November 7, 1994

"This fucking thing is fucking fucked."

Steve Gurney
Satellite engineer
December 1994

"We believe these launches have the same import as the innovation of the telephone and the personal computer. For the first time, individuals will be able to communicate with other individuals anyplace in the world, at any time."

Gary J. Reich
Prudential Securities, Inc.
January 1995

"It's too bad that software doesn't have a higher priority. Here they just don't recognize software as that important. In the end, it's become this free-for-all, with a lot of egos getting in the way. I don't know, it's gotten really tenuous. A lot of things are right on the edge of not working."

—Steve Gurney
On his last day at Orbital
February 1995

Krebs Board

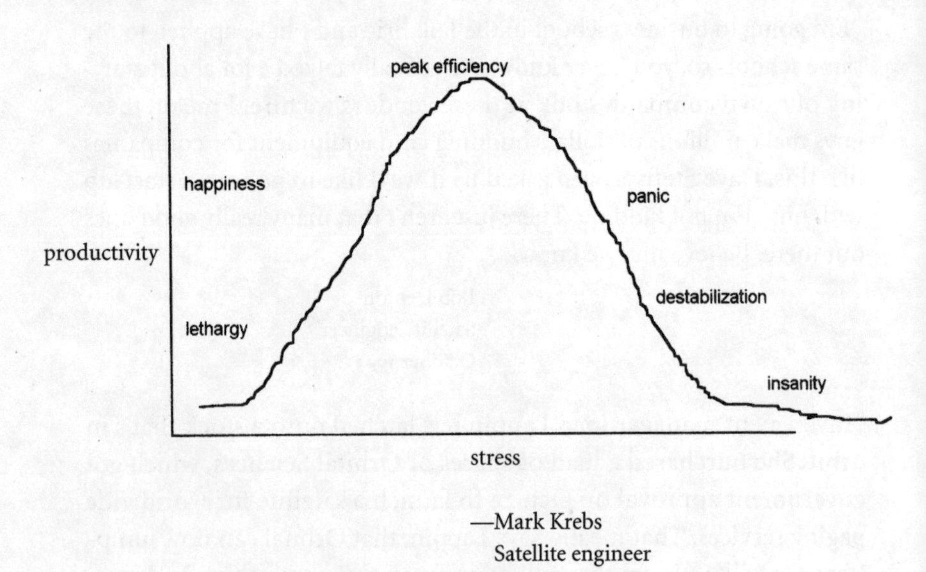

—Mark Krebs
Satellite engineer
February 1995

"Sure, it would be better to wait a week or a month if there were really a problem, and incur that much more financial loss rather than taking a death blow of mission failure. But it's not financial pressure that's forcing this launch. The baby has been worried over long enough, so it's just time to launch it now. Ready or not. Right now."

—Mark Krebs
Flight journal
March 1995

Fortune cookie predictions shaped into a fan

Found in Grace Chang's office a week before launch

1. You will take a trip to the desert.
2. You constantly strive for self-improvement.
3. The real problem with your leisure time is how to keep other people from using it.
4. Your dreams will soon come true.
5. You will soon enter a new line of work.

"Launching a satellite is not a process that can be rushed. Intelligence officers will tell you that almost every instance in which a reconnaissance sat is launched in a rush, it ends with failure."

—Jeffrey T. Richelson
America's Secret Eyes in Space

"I had a dream the day we were going to ship the satellites to Vandenberg. I woke up in the morning and went downstairs to the high bay to tell everybody to do some tests. And for some reason as soon as I walked into the high bay, Paul looked at me and said, 'My God, we forgot the brakes!' I was like, 'Oh my god, we forgot the brakes!' Then he said, "Dave says we're gonna have to do it in software." I woke up immediately and thought, 'What brakes?' Something horrible was about to happen. I just knew it."

—Annette Mirantes
Satellite engineer
March 1995

CHAPTER EIGHTEEN

March 1995

Flounder, Arnold, & Vlad

Sixty years ago Burton Mesa was a quiet, lonely, strangely beautiful plateau at the edge of the Pacific Ocean, a dry, windswept stretch of Santa Barbara County between the famous flower fields of Lompoc and Santa Maria townships. For a long time the mesa served as a place for cattle and sheep to graze over sparse vegetation. The most notable activity occurred at night, when great, quick-moving fogs rolled across the flat land and sponged their way into the hills.

In 1957, as home of the U.S. Strategic Air Command, the mesa became a training field for military exercises, a bed for missile launchers, fuel storage tanks, power houses, blast-proof control blockhouses, radio guidance stations, high bays, 20-ton cranes, liquid oxygen–generating plants, and telemetering stations. As birthplace of the American's first ICBM, the mesa's flat shale landscape was overtaken by 135-foot-tall gantries that looked like oil derricks and rockets capable of hurling nuclear warheads 5,000 miles across the ocean, all standing visible in preparation for war.

In 1961, then known as the home of Vandenberg Air Force Base, more than 18,000 people lived on Burton Mesa, including employees from several of the nation's primary rocket manufacturers, such as Douglas, Martin, and Boeing. They were joined by a stream of technical and support personnel from the navy and air force, and their families. Boy Scout troops sprang up, as well as women's clubs, Little Leagues, schools, and churches, making the base and the little towns of Lompoc and Santa Maria centers of domestic activity, as lively and all-American as any place in the country.

Those who saw the first launches said rockets flew like silver bullets straight from the ground. A ball of flame, tail afire, an Agena or Discoverer rocket blasted into space carrying many of the first American satellites into low-Earth orbit. Vandenberg rockets hauled heavy packages of science instruments that exploded from the rockets' third stage and sent satellite capsules plunging wildly back into the Earth's atmosphere northeast of Hawaii.

During the 1980s engineers from Vandenberg launched at least forty rockets a year. But by the early 1990s, in the wake of the cold war, the place had quieted down considerably until the Defense Department launched only ten to twenty rockets a year there. The government had awakened to commercial possibilities, however, and the State of California decided to support the development of what some predicted would become a burgeoning commercial spaceport. Aerospace executives and federal officials talked about making a mutual effort to convert excess air force launch sites to nonmilitary use, and Vandenberg was one of the first.

The vision that David Thompson and a few others had of a robust commercial space industry showed signs of becoming a reality by 1995. After the Persian Gulf War sales of GPS receivers started to grow at double-digit rates. PanAmSat, the commercial rival to Intelsat, took advantage of the rapid expansion of global television and quickly built a global network of four satellites to offer coverage of seven continents. Sales of private-network satellite communications in China were booming. A business called DirecTV opened its doors in the United States offering high-power digital, direct-to-home satellite television, and rapidly attracted close to a million subscribers.

Privatization, competition, and a quantum leap in satellite technology drove the boom. Spot beam antennas worked at greater accuracies using less power. Batteries were more efficient. Solar panels were more efficient. There were 145 commerical satellites in orbit, with another two dozen planned by the end of the year, and a hundred more on order. The International Telecommunications Union saw people clamoring for more bandwidth and counted more than 900 applications on file for new services.

Driven by business projections for hundreds of satellite launches over the next ten years, the entire industry shifted focus from military projects, which were already dwindling, to commercial prospects for new launch vehicles, Earth stations, spaceports, and ancillary equipment and services. Space industry rhetoric that once flourished on words like "free society . . . superpower status . . . public opinion . . . Cold War," suddenly awoke to "free market . . . investment financing strategy . . . mass market . . . commercial products." Over the next ten years, it was projected, there would be 273 commercial geosynchronous satellites launched, with a value of $37.8 billion.

Then, of course, there was the Klondike in low-Earth orbit. Motorola kept raising millions of dollars for its sixty-six-satellite Iridium system. Teledesic, supported by Bill Gates, had started hiring engineers to staff its first development teams for a megaconstellation of more than eight hundred satellites. Other businesses called Globalstar, Odyssey, Gonets, Celestri, Ellipso, and Starsys expected to enter the battle with different configurations, and many were well on their way toward building satellites and implementing a business plan.

Wall Street responded, and the aerospace business began to reinvent itself.

So in 1995, hoping to kick-start the business in California, the federal government had just started offering the Vandenberg base to commercial space companies, promising a cut rate. Orbital Sciences actually was one of the first to acquire its piece of the historic mesa not far from the pads where many of the first American satellites were launched, within just a few miles of where Minuteman missiles still broke from underground silos for regular target practice in the Pacific. The feds hoped to revive the mesa to accommodate a growing business, offering the site for one-quarter of its former costs, but when Dave Steffy's engineers started flying in from Washington and driving out to the base in their rented cars and vans, they found the area oddly barren, almost devoid of activity.

In March 1995, fresh from a painful testing program, the satellite team found a temporary home in old Building 1555, a high bay and production facility at the edge of the air force base. Young engineers found themselves walking, day and night, across empty plots of ground that sometimes seemed as desolate as a desert island.

Dr. Strangelove may have withdrawn into retirement under the eighteenth green of Vandenberg's exquisite golf course, but heavy fog still rolled across the mesa at night with unpredictable regularity, mysterious and ominous, claiming the landscape as its own.

The team felt comfortable there, at first. But soon, particularly late at night when the rains came and winds off the ocean rattled the roofs of the old high bay and whistled through the rafters, the final destination of their journey turned surreal.

And when the rains continued and the fog rolled in to stay, Burton Mesa seemed again like a strangely secluded place, an isolated landing on the team's last leg of a long and difficult pilgrimage.

March 5, 1995, Sunday

Vandenberg: green and wet.

Heavy rains swept in, causing floods and mudslides from Lompoc to Santa Maria. Dave Steffy drove up from San Diego where he met with a crowd of consultants. Useless now that his engineers had taken over in Building 1555, Steffy spent the day with a book titled *Digital Signal Processing*.

The rest of the engineers, who arrived with the satellites, operated solely on adrenaline as they continued forcing more and more limited performance tests on FM-1 and FM-2. Rob Denton called Annette back at Steeplechase and asked her to rewrite a piece of his power system's code.

"That gives me a total stomachache," Annette said, but then she agreed.

The Pegasus rocket, stretched out on the high bay floor, gleamed with a new paint job, but without decals it looked as plain as one huge log of PVC pipe. A fight broke out over which logo would appear at the top of the rocket, Orbital's or Orbcomm's or NASA's, and although it struck everyone as a particularly absurd reason for conflict, it actually required delicate negotiations to resolve.

A busload of Orbital's vice presidents was coming out in two days to "kick the tires and get in the way," according to Steffy.

At eight P.M., Steffy and his assistant, John Stolte, were sitting in a fast-food joint near the motel drinking a couple of Sam Adamses when Steffy mentioned that he had just called his wife, Jennifer, and detected a note of anger in her voice. She didn't say she was sitting at home by herself with a sick kid—again—but Steffy felt guilty, anyway.

As more engineers arrived from Virginia, the pilots on the local puddle-jumper, *Mesa Express*, told people they had followed Orbital's stock price for a couple of years. They waited for someone to say anything about the launch, but nobody ever did. The stock had stalled around 17, and aerospace analysts on Wall Street predicted it would shoot to 50, if these first two satellites went well.

Of course, Wall Street analysts had not been in the high bay recently.

March 6, 1995, Monday

No meetings, no briefings, no conference calls. The engineers plunged into final tests.

I saw Steffy wander around the building early to set up an office and finish his last-minute paperwork. He told Stolte, who had been supervising since three A.M., that he would spell him. "Go back to the motel and get some rest."

Of course, Stolte didn't want to leave the satellites either, and he only laughed to himself when Steffy slumped away. "They're gonna have to peel Dave away from here," he said. "The most miserable time for any engineer is when they've finally worked themselves out of a job and there's nothing left to worry about. Looks even worse for a program manager."

Warren Frick, the Pegasus program manager, came in about nine this morning and stormed into Steffy's temporary office, angry because forty satellite engineers had taken over the high bay where he planned to mate Pegasus with the satellites. Steffy's engineers, he complained, refused to follow rules.

Warren blustered around until Steffy finally agreed to talk to his team. Warren insisted that everyone take thirty minutes out of their day to watch a safety film—a mild form of retribution.

Warren also let it slip that Pegasus was having problems with its PVCs and windings, some of which were apparently broken. He rested his head up against the wall while Steffy tried to diagnose the problem in detail—a counter form of retribution.

Around noon the man who headed up Orbital's operations at Vandenberg, Ed Rukowski, walked in and mentioned that he had done pretty well with a microbrewery business he opened with a buddy in Virginia. They had started distributing a beer called Potomac.

"Now there's an idea," Steffy said. "Give people what they really want. Not global communication—forget rockets and satellites. Sell 'em beer!"

No joke. Ed said, in fact, that in a few years he expected to make enough money from the beer business to retire. "I'm moving to Montana to set up the next microbrewery," he said, "and after that I'll just sit back and watch the profits roll in."

By midday, Steffy finished his fifth cup of coffee and was so jittery, he spent a considerable amount of time compulsively arranging a set of fans to cool his office. There was really nothing else for him to do.

At seven-thirty, I found most of the engineers lolling around in the hot tub outside the Embassy Suites motel, after enjoying a languorous happy hour, eating popcorn and drinking beer. John Stolte grabbed his paunchy stomach: "Look at this baby! Oh, yeah!"

They talked about how much weight they'd gained over the course of the project. They talked about Denton's nerdy bathing suit, about wives and babies and what they wanted for supper. Some of them wanted only to get drunk. One of the technicians, Alan Murray, showed up with a mean burn from having spent the afternoon sunning on the dunes.

Mostly they worried. In the first satellite pass after launch, would the satellite make the orbit right side up? Even though it will be tumbling, it should be tumbling slowly, Steffy said. Denton argued there was no way. He gave a fifty-fifty chance that the satellites wouldn't pick up a signal during the first few orbits. He also tried to convince Steffy they should start the party as soon as the baby dropped. Steffy said no.

People swamped the hot tub: technicians, program managers, engineers—sweating together, bubbling in their own juices like boiled lobsters.

It seemed strange that everything had gone so well. The hard work was almost over.

"Our work ended two weeks ago," Denton said.

"No," Steffy said. "I'd take one last look, if I were you."

But even Tony Austin, the Q&A guy, hadn't found anything to complain about. The next day at one, they would start a series of launch simulations, and after that, party time.

Everyone seemed eager for Mark Krebs to arrive. He would be spinning. They laughed just thinking about Krebs, who stayed so damned nervous most of the time now.

The freakin' antenna was modulated well enough, Denton decided. It would probably send signals even if the spacecraft flipped upside down.

"Better call the wife," Alan Murray, the technician, said. He stepped out of the swirling water.

"Does anybody have beer in their refrigerator?" Denton asked. He also slipped out of the hot tub to cool off in the swimming pool.

Someone started to rank on Grace Chang. Apparently she had tried to make one of the technicians move some equipment for her in the high bay that afternoon.

"I told her to move it herself. She's a real—"

No, no, no. Everybody defended Grace. She was just not a passive woman, that's all.

The stars were just coming out. The engineers climbed out of the pool, reaching for their towels, and agreed on a time to meet for supper.

After dinner they returned to the high bay and worked through the night.

March 7, 1995, Tuesday

The scene around the bay was mellow. After long hours but few headaches, the team decided that all the work they did before leaving Steeplechase had paid off.

John Stolte, who came in again at three A.M., slumped back to the hotel at noon to sleep. Steffy was still walking around with his hands in his pockets with nothing to do, while the rest of the team whispered about him behind his back. In his absence, most of them said, things went smoothly.

For no apparent reason, the engineers started speaking in German accents. They played CDs all day on a box Stolte brought for them: *Tom Petty, 10,000 Maniacs, Weezer, Portishead, Green Day, Soundgarden* . . . Grace sang along at the top of her lungs. So it had been, day and night.

The engineers also developed a kind of weird emotional attachment to the satellites. They talked about the three spacecraft as if they were prodigious children and named the satellites, inexplicably, *Flounder, Arnold,* and *Vlad,* which they pronounced in German: "Vlown-dah . . . Ahnald . . . Vlhot."

The long hours accrued since Saturday—sixteen to twenty a day—seemed like nothing more than a way to avoid separation anxiety.

When they took a break for an afternoon meal of dry raisin bran and vitamin C , Grace yelled, "We look so pasty!"

People walked around with amused expressions as engineers frantically typed out test procedures just moments before conducting final tests.

The increasingly crazy, slapdash, free-for-all, unprogrammed approach to engineering—they claimed—was working. Of course, one of them had an ulcer. Grace's stomach grumbled. Every one of them looked greasy and tortured.

The simulation that was to begin at one P.M. did not start until after ten P.M. because of a few programming errors, and then the Pegasus rocket failed to reach orbit. No one knew why the rocket, carrying the three satellites, crashed, but as far as anyone could tell, the simulated rocket was left bobbing in a million simulated pieces out in the middle of the simulated Pacific Ocean.

Kim Kubota blamed Dave Steffy. Steffy complained privately about Kim's outbursts and reminded Stolte that she needed to take some time off. But no one could make Kim stay away from the spacecraft. She refused to leave the high bay.

The last few months had been intolerable for Kim. As the one person in charge of the entire testing program, only she knew just how close they were to the edge. One day she told me that she used to love Orbital's philosophy of using small teams to build spacecraft. Today, she said, she had changed her mind: "There's something to be said for being a faceless engineer."

March 8, 1995, Wednesday

Last night's investigation of Launch Sim 3 showed that Pegasus had not crashed in the ocean after all. In fact, it nailed its orbit. They traced the error to a software glitch.

However, the satellite team discovered that a few hours after the satellites were ejected into space, both Orbcomm satellites lost control of their flight computers. The engineers stormed around saying it was something they had seen two weeks ago at the Steeplechase lab and Schoen's engineers chose not to fix it.

Kim insisted on repairing the problem immediately so the Losers Club could take a day off. The team wanted to do some shopping in Santa Barbara.

It's funny, the old technicians around here said, but the public never heard what went on behind the scenes. Like the time a bunch of engineers at Fairchild dropped Topex, a two-ton satellite, from 22 feet off the ground because the numbnut in charge decided to lift it up during its last tests with cables that hadn't been gauged. Then it took eight weeks to test the cables and restring them.

No one heard about time they were moving a $35 million Defense Military Satellite down 13th Street on the base, making the turn onto New Mexico and knocked out a stoplight.

Not many people know the sound of anxious silence filling a high bay when someone makes a mistake so significant that a multimillion-dollar federal contract goes up in flames in an instant.

Why, I wondered, were they talking about such things now?

March 9, 1995, Thursday

The satellites became a proprietary matter today. FM-1 was Grace's vehicle, called *Flounder*. FM-2, which was Kirk Stromberg's, known as *Vlad*. Paul Stach claimed the Qual, *Arnold*.

To these three engineers, the spacecraft took on lifelike qualities, like embryos, and they, the engineers, claimed they were responsible for preparing the way to birth. Combing their monitors for errors, grooming bit streams, consolidating test plans, identifying the final errors and fixing them, they stated they would not stop work until the final moment of deliverance.

But why were they so angry?

The routine made them insane: in at work at seven A.M., not out until midnight or sometimes three A.M.; fingers flying over keyboards, staring interminably into glassy monitors. Every one of them turned phobic, afraid of catching a cold, coming down with the flu, passing germs. The only thing that could have stopped them, they said, was illness.

None of the engineers had a two-year-old at home to call, no husband or wife wondering about their whereabouts. They fell spellbound in the high bay, uncommunicative, beyond reach; with their false accents, cracking inside, they seemed lost to their own warp and woof.

"It's sick," said John Tandler, the satellite watchdog from Orbcomm Inc., as he saw them take over one computer station after another and command sets of test racks.

And yet every one of them had been there at some point in the project—horrible hours, no life, hating their jobs.

They had fallen in love with a machine.

Today the rumor was that Mike Dobbles wanted to come back to Orbital and work on the company's new reusable launch vehicle. They also heard that back in Virginia engineers on all the projects had been ordered to work sixty hours a week—a reduction for Orbcomm's engineers, but not a nice deal for everyone at home.

In the kitchen I heard Kim Kubota laughing at the rumors. This is the life to which they had been resigned, she said. This kind of behavior was exactly why she wanted to leave Orbital, she said. The affair's gone on too long, she said. They bought her; they sold her. Now she wanted out.

She said all that, drank her cup of water, and marched right back through the big doors to conduct another test.

March 10, 1995, Friday

High winds had whipped up off the ocean from a terrible storm the night before last, slamming the California coastline. A gust ripped part of the roof off the east high bay, next to where the Orbcomm and Pegasus teams were working. The large doors leading from outside into the bay rattled like timpanis. The roof shook, winds whistled through. Outside by morning a couple hundred yards away down the rocky coastline, water was visible. The surf clawed at the beach, water streamed inland, leaving roads awash in a flood.

The engineers finally finished Flight Sim 4 late last night and left for Santa Barbara. News reports said bridges and parts of some major roads were unpassable, sections of highway had disappeared as boulders tumbled and mud slathered the coastline.

Left to himself in the high bay, Steffy started troubleshooting the garbled telemetry that came in on FM-1 last night. He had Annette on the phone from Steeplechase and put in calls to a couple of technicians to help him debug. Watching data streams, trying to catch the glitch, he looked unusually satisfied.

He didn't complain that the other engineers had left.

"It's unfortunate," he said, but he was not convincing.

New problems with Pegasus showed that the rocket's batteries may have died, a problem that would set back the schedule a couple of days.

Tony Robinson finally arrived and started practicing the delicate art of folding his antenna into the Qual satellite's narrow trough. His wife showed him how to tie the ribbon before he left home. People around the bay gave odds that once the satellite was stowed and mated, the launch would bump.

If rain delayed them for more than seven days, the antenna would lose its springiness and suffer what they called a "hysteresis effect," meaning it would come out of round. If the antenna deployed as an oval, the satellites would begin life with birth defects akin to a cleft palate and ruptured eardrums. Most people set the odds at fifty-fifty, which wasn't bad, Tony decided, considering.

The rumor today was that the insurance company covering Orbital's launch and satellites had agreed to insure only one satellite. Take your pick which one. The company's lawyers had been arguing the issue for months, but no one could change an actuarial table. Aerospace remained a high-stakes gamble everywhere.

Since there were so few engineers around, the rumor passed and died quickly. Nobody left in the high bay wanted to take the time to think what that meant the sane world thought about their chances for success.

March 11, 1995, Saturday

At about seven o'clock tonight, while mating Microlab, the engineers heard a loud pop.

They demated, cleaned out all the connectors, remated, and never heard the sound again.

You never know, they said. It's not always rocket science.

March 12, 1995 Sunday

"Mating? What could go wrong? I mean, how long does it take to mate thirteen male connectors to thirteen females?"

Everybody was in the hot tub last night when one of the techs started asking questions that sounded like the joke of the day.

But this morning they were all business. Pegasus still suffered from the same obscure battery problems, and some members of the satellite team hadn't been heard from since they left for Santa Barbara.

News reports from the region told of major problems around La Conchita, where townspeople described seeing not only mudslides but whole slabs of mountainsides tumbling off into the ocean and slices of the highway disappearing into the Earth.

While the Pegasus rocket team continued troubleshooting, Grace Chang, leader of the Santa Barbara expedition, suddenly showed up in the high bay kitchen and began flipping through the Land's End catalog, looking at shirts, hats, and sweaters, which Stolte had ordered for Orbcomm engineers.

Grace was wearing a rusted brown sweatshirt with a pointed hood, which she had purchased at the Santa Barbara K-mart. With the hood up, she looked like Yoda. She seemed to enjoy telling stories about the team's adventure in the storm.

"We were homeless," she said.

Their shopping spree was especially rewarding because no one else in the area would risk going downtown during the storms. Sopping wet, a few of them actually left early and ended up dodging boulders on the highway back to Lompoc. But Grace, who claimed she was unable to find a station on her car radio that gave weather information, stayed in town for the night. She drove to the local Holiday Inn, found people sleeping in the lobby, and was turned away. She went to get a toothbrush and toothpaste from a twenty-four-hour convenience store and ended up hiding behind a head-high stack of bread to brush her teeth that morning.

A lively raconteur, Grace kept people entertained in the high bay kitchen with her tales.

"God, K-mart!" she moaned. "It was awful! You wouldn't believe how much time we spent just looking for the right thing to wear. I mean, if you have to buy K-mart clothes, you at least have to find the right colors or something."

Having survived the flood, most of the engineers actually seemed happy again.

Out in the high bay, Mickey Doherty, the team's salty Irish technician, kept calling Tony "homey" while they tried mating the two Orbcomm satellites through the evening. Having an Irishman call him "homey" was not acceptable to Tony, who was especially proud of his English heritage. Mickey also went around the high bay teasing Tony about his accent, talking about "rubbah snubbahs," devices Tony designed to cushion the satellites.

But as they fought, neither one of them could say "rubber snubbers" very distinctly, a fact that made their feud even funnier to everyone awake enough to observe.

March 13, 1995, Monday

The launch was delayed by one day, due to the Pegasus battery problems.

Microlab also had troubles—one pin connection wasn't quite firm, according to the technician who accidentally discovered the problem this morning during the satellite's mate to the rocket. If he hadn't taken a moment to wiggle the connection with his hand, it would have demated during the launch.

They spent several hours demating and examining the satellite, and then someone decided they should actually fly in experts from the company who manufactured the cables and let them do it themselves.

Mating a satellite was not easy. A minor oversight—one loose connection—they're dead in space. As with Orbcomm-X, no one would ever know what happened.

Out to dinner: The engineers squeezed into a couple of rental cars and sped north to a steak house for prime rib. They ordered a couple of brands of Scotch and a 1988 Pinot Noir to heighten the taste of the meat, the best corn-fed beef any of them had eaten in years.

They railed against Steffy before the drinks arrived, and then a few of them planned a day of golf at the air force base.

Krebs confessed that he worried about what he would do next—a topic that kept coming up during dinner and finally killed the conversation completely.

"What do you think the best strategy is?" he asked around the table. "Just wait out this launch until they finally come begging for you to join another project?"

The other engineers thought he was crazy to worry. Some people thought he was one of the most eligible management prospects in the division.

But then everybody had seen engineers promoted who were the stupidest boneheads they ever worked with.

As they talked, they became more and more critical of Orbital's management. They worried mostly about what would happen once the team moved to Germantown.

After dinner, back at the motel, they climbed into their swimming trunks and skivvies and slipped outside into the Jacuzzi.

Eric announced that he was leaving for Stanford to get his MBA.

Tony Austin showed up with a six-pack of Lite beer, and everybody started drinking again.

Krebs kept submerging himself in the 110-degree water, then finally came up to cool off for a few minutes and said: "You know, the only one here who really believes everything's going to turn out okay is Dave Steffy."

"He's also saying there's no way that Germantown will take over our business," Denton observed.

Poor Steffy, they concluded. Brainwashed. Corporate weenie. A lifer.

March 14, 1995, Tuesday

Late in the afternoon, two of the engineers sneaked out of the high bay, drove back to the motel, and slipped into the hot tub. Ed Nicastri, the new head of the Space Systems Division, wandered in unexpectedly—they could hear him jangling his keys—and a couple of Orbital's other big brass were tagging along, too. They spied Krebs and Denton soaking up the last hours of sunlight in the tub.

Krebs disappeared under the water. Denton moved to the opposite end of the hot tub as soon as he realized they were headed his way.

Big Bill Kauffman, the program manager for Microlab, walked outside then, too, looking sour and tired.

"Why aren't you happy?" Krebs asked.

Bill growled, "I'm happy," then headed straight for the bar, ordered a double bourbon, and slouched back to the Jacuzzi.

"Heard any new rumors?" Krebs asked, and Bill echoed, "Have you heard any new rumors?" For a moment they were at a standstill.

Finally, Bill said, "I always have to check out every rumor I hear, and unfortunately—or fortunately—they're always about 50 percent true."

Krebs said he appreciated the fact that Bill bothered to calculate the probabilities.

The engineers wanted to be left alone, without management supervision or attention. At least for one evening, they wanted to stop thinking about technical questions and company politics and any little discrepancy in their test logs.

But the bosses always wanted to hear—for the *n*th time—from Krebs about attitude control or magnetics, and Krebs would always have to put aside his doubts and say everything was great. What else could he say at that point?

And Steffy would want to know what would happen if the Pegasus launched the satellite at a 75-degree inclination, and Krebs would say, "I don't know, I never tested it at 75 degrees," which would be just another little joke, since Krebs had tested every angle between 60 and 90 degrees a thousand times in the past two months, and Steffy knew it.

So tonight Tony Robinson called them all to the Thai restaurant to eat and drink, and by ten-thirty, an hour and a half before the night's Minuteman launch, they finally became so ridiculous that someone started a food fight at their table for twelve.

After three hard years of work, the satellite-builders thought they deserved an evening free, without the press of management around the table or the boredom of office politics to inhibit the flow of unfettered conversation.

A little after eleven o'clock, once the food fight ended, the Thai waitresses were saying to the engineers, "Go home, go home," because they wanted to close the restaurant.

But the satellite team did not seem to hear them because the engineers were too boisterous. It was not that they meant to be obnoxious—they were just noisy—but no one could understand the constant, terrible pressure to perform at consistent peaks for two years, to hit a sequence of impossible deadlines. The Thai waitresses couldn't possibly have known or even understood—no one would have had any idea, really. No one.

So the Thai waitresses kept telling the satellite engineers, "Go home, go home," while some of the men kept running back to the bathroom, where there was a little statue of the Buddha by the toilet inviting anyone to leave a few coins and make a wish, which some of them did, several times.

Finally, Tony took the check and handed it to Stolte. The team raked up a $20 tip, and the Thai women sang: "You will come back tomorrow night, you will come back again Saturday night, yes yes yes?"

"Yes, yes, yes?" they sang.

After the launch?

The engineers stumbled out to their cars.

And this was where they went: to the hotel and to bed, or flying over speed bumps on the air force base to reach the high bay before it was too late for another look at their babies, then back to the Jacuzzi, where they would try to turn up the heat high enough to sweat out all the beer and be ready to rush out in the morning at three to work again.

But it was really over for them. There would be nothing left to do tomorrow afternoon.

Then it would begin like clockwork, the final closeout. They could start asking important questions, getting answers to rumors about who had been doing what to whom in Germantown, and who would have a job after the launch, who would get a promotion, who would be laid off, and what would happen to the team.

John Stolte and two technicians, Mickey Doherty and Alan Murray, were the only ones left in the Jacuzzi after midnight, still steaming off the last ounces of beer, putting the day behind, when Mickey admitted that he had also been sweating earlier in the day. His palms were soaked during the final mate of the satellites.

"I looked down into that black hole and—"

Stolte said: "And you did real well, Mickey. I told you it would be good."

"Yeah, but I was scared I'd drop the friggin' thing and the—"

"The snubbers—the 'snubbahs' wouldn't—"

"Yeah, I was lookin' into Tony's 'snubbahs' and I could hear Tony up there talking at me—"

"But Chip was doing a good job, wasn't he? He was keeping quiet, just like I told you," Stolte said.

"Yeah, I was proud of Chip."

"You know, he got a sick stomach right before the mate."

"Nerves," Stolte says.

"Damn right, nerves!" Mickey said.

"Yeah, and I saw you rubbing your hands, but I didn't know what the deal was."

"I was cleaning the sweat off my palms, mate, that's what the hell I was doing."

"And you did a great job, too, Mick."

"I couldn't even sleep the last two nights thinking about it. I was up all night—twice! My mind's racing!"

"I heard the Microlab guys say they couldn't sleep either, because their program manager kept climbing in and out of bed all night going to the bathroom."

"You'd never suspect that now, would you?"

"No, they're pretty cool around the high bay."

"I was looking down into the darkness waiting for that satellite to settle in and make the connection, and I'm thinking to myself, 'Mother Mary . . . '"

"That reminds me, Mickey, you've got off Saint Patty's Day."

"Can't do it, John. Gotta work. Our guys need all the help they can get."

"Well, I'm going home, so I'll leave it to you. Dave Steffy and I have to catch a flight about nine."

"Oh, good God! You're flying back with him?"

"No, you got it wrong, he's cool. I think he's finally chilled out."

"Yeah, you think he'll be fine tomorrow?"

"Better be, there's only two days left and we launch."

They boiled in the Jacuzzi until someone at the motel finally turned it off. Stolte was up immediately, popped out of the water, and raced over to find the switch to clip it back on. He had become like a mascot to the engineers, always wanting to keep them happy. And don't think they didn't appreciate it. They knew who took care of them.

"No, you guys did a beautiful job today," Stolte said. "You deserve some time to unwind."

Of course, Alan Parker's guys would complain that the satellite-builders were still burping beer when they came back to the high bay at three in the morning.

They would come in to demate the satellites again—pin three on the umbilical connecting Orbcomm to Pegasus had failed—and a whole new round of troubleshooting would begin.

Burping beer and stone-cold sober.

So that was it at four, as they demated and hoisted FM-1 carefully up in the air with a pulley, a team of technicians wearing rubber gloves, using small mirrors and tiny silver implements that looked like dental tools, squinted into the closed, dark body of FM-2 to check the play in every connection. They found one with a couple of hundred thousandths of a wiggle.

"We fix it now," Tony said.

"Your guys are tired," Stolte argued. "Can't it wait until the morning?"

"We're used to hard work," Tony answered. "We wait till the morning, we've got Pegasus engineers and half of Orbital's brass here. We get it done before the circus starts."

Mickey walked up, eyes bleary but grinning. "Go to bed, homey. You come back in the mornin' and these satellites will be off on their honeymoon."

Tony argued awhile, just to be a gamer, then said: "Well, just keep management out of here," and after twenty-two hours at work, he wandered out to get some sleep.

March 15, 1995, Wednesday

"What are we going to do with a free day?" an engineer asked at breakfast.

Stolte had left for Virginia. Steffy had vanished.

The final flight simulation had gone so well that no one else was called out of bed. The connector problem turned out to be a false alarm. Even Ed Nicastri was walking around aimlessly. With nothing to do, he finally quit jangling his keys.

Before the final closeout the engineers inscribed the names *Flounder* and *Vlad* on a decal and secretly planted them on the back of the payload shelf near the solar array drives on both satellites—one final act of defiance against management—and claimed the technology as their own.

The sky looked clear around Lompoc. The weather was fine, 70 degrees and breezy. The sign at the Embassy Suites that once said, "We Welcome You, Orbital Sciences," had been changed to "We Welcome You, Central Coast Convention of Dental Hygienists."

The young satellite builders—those who remained, anyway—were full-fledged commercial engineers at last. Delighted by news of newborn babies, low mortgage rates, vacation plans to the Bahamas, and any jump in the stock price, they were prepared to spin out into orbits in a profession that remained uncertain and risky.

But surely, they believed, they had accomplished something astonishing. At the moment, perhaps, they couldn't imagine what could be more challenging or, potentially, more profitable.

If Dr. Strangelove had retired and were still living somewhere underground here amid the silos or hiding among the cypress and eucalyptus groves, he was a man certainly out of the loop now.

In a few days, they would know. Life would be good again. They would beat all of the competition into orbit.

MARK KREBS: So the stock went up two dollars this morning.

MARIA EVANS: Who cares? We don't even have contact with the satellites yet.

MARK KREBS: Nobody knows that. The press release just said "successful launch."

MARIA EVANS: Well, I'm going to wait until we launch the constellation.

—Conversation
Launch day
April 3, 1995

April 24, 1995, SPACE NEWS

Orbcomm Gets Scrutiny
Teleglobe Weighs $70 Million Investment

By Patrick Seitz
Space News Staff Writer

WASHINGTON—Orbital Sciences Corp. officials are scrambling to rescue two flawed Orbcomm communication satellites, complicating a pending decision by a major investor over whether to pump more money into the revolutionary global messaging system.

The two satellites were supposed to be the first cluster of an eventual 26-satellite low-Earth-orbiting (LEO) constellation. Both satellites launched April 3 on a Pegasus rocket have suffered anomalies that prevent them from communicating with ground equipment.

Engineers discovered a problem April 15 in the subscriber communications subsystem on Orbcomm 1 satellite, adding to headaches caused by a previously reported problem with Orbcomm 2. Shortly after launch, Orbcomm 2 developed a problem with its gateway communications subsystem, preventing the satellite from responding to transmissions from ground stations. . . .

"Who put those 'McDonald's Is Hiring' signs up everywhere?"

—Annette Mirantes,
April 1995

"It's tough to be a pioneer. Ask the folks at Orbital Sciences Corp. They're trying to ring the globe with 26 communications satellites that will carry data and other kinds of information to customers around the world. But like the pioneers who went west, the Orbital scientists keep getting stuck in the mud."

—*The Washington Post*
April 24, 1995

"The stock is dead in the water unless they fix the problems."

—Otis T. Bradley,
Aerospace analyst
Gilford Securities

"They're paying the price for being a leader in this endeavor."

—Charles M. Robins
Analyst
Pennsylvania Merchant Group Ltd.

CHAPTER NINETEEN

April/May 1995

I_AM_DESPERATE

Two circles slipped across the conical projection map with unbearable lightness, passing from Portugal to Cape Verde to Mombasa and Calcutta, from Seville to Cape Horn to the Spice Islands, pulsing steadily on their paths above the Earth's surface, moving with the effortless efficiency of long-distance runners.

The hands of operators flashed across mouse pads like chess masters in competition. Above their heads a projector cast a bright color image of the planet overlaid with screens from the engineers' monitors against the far wall. In the dim light telemetry gauges blinked red, yellow, and green, while black windows of agatelike code telescoped into view. As the engineers magnified data pouring in from distant Earth stations and reviewed error logs for clues, every move they made played across the far wall for the benefit of spectators shifting outside the periphery of the control room like witnesses at an execution.

"Just show me the accounts receivable screen," muttered the Orbcomm business manager, pressing his fingers against the glass, squinting into the room. Other executives crowded around, peering in as each new screen emerged, then cursed or mumbled as it vanished in a blink. Raw satellite data was, to many eyes, indecipherable.

David Thompson turned his back and leaned against the glass. His thin fingers riffled pages listing the estimated times of satellite passage above Earth stations in New York and Arizona and scanned lists to find the next scheduled pass. He had grown impatient with the process but, with nothing to do now but watch, he had merely resigned himself to standing uncomfortably outside the control center at all hours of the day and night, though he refused to let up in his demand for answers. Perhaps praying—or at least hoping against hope—he anticipated the moment when his operators would reset the paralyzed satellites and restore his business before it collapsed around him like a hail of meteors, leaving a billion-dollar market just beyond his grasp.

Annette Mirantes tugged the George Mason ballcap down lower on her forehead until it almost touched her eyebrows. She set her teeth and glared into her monitor.

"Somebody pull the curtain!" she said. She angled her chair so more of her back faced the crowd on the other side of the glass. "Don't those people have jobs to do?"

The engineer beside her said nothing, merely tapped his keyboard to scroll through another dense block of code.

Annette glanced over her shoulder to see how many were watching them now. She could tell they were talking, but she could not make out the words.

Her supervisor cupped his hand over the receiver of the telephone. "There won't be jobs for any of us to do if we don't make this thing work," he said hotly.

She had expected to be getting drunk by now instead of swilling Evian water and nibbling raw vegetables in the network control center. Sleeping on the floor of her office for the past week, Annette had at least quit having the hairy nightmares that plagued her for the past year. The daily succession of graveyard shifts, although exhausting, did heighten the sense of numbness she experienced now and provided her some relief from disturbing dreams.

Rumors that one of the enraged executives had hurled a fax machine across his office, that the value of the company's stock teetered on a precipice, that the public relations staff could not forestall the media much longer, came and went with every shift change. She had found it easier to ignore them as the days passed, though she had yet to grow accustomed to the distraught expressions of the tall white-faced men in suits who lined up along the glass wall and watched them work, unsuccessfully, for the past ten days.

Annette tapped her mouse and closed a window containing the flight computer code. With an hour until the next satellite pass, she pushed away from the console and took a deep breath. She could afford to take a break now and review some of the back-orbit data collected during the past week.

When she stood, she saw them again, leaving smudge prints against the glass. She then reached for the silver handle and leaned into the heavy door. The air, so dry and cool, seemed to whisper as she stepped out.

Across the hallway in the conference room, several operators from the previous shift had gathered to sort through long stacks of accordion paper thick with telemetry from the most recent back-orbit dump. On a corner folding table, the remains of a ten-day-old celebration cake lay in a heap of crumbs and chunks like dried spackle.

Annette pitched her cap into a chair along the series of third-floor windows and fluffed the crown of her thick red hair, letting it spill down her back. She bent over to rub the moisture from her palms onto her sweatpants and then rose in a patch of sunlight that broke warmly through the windows. At least the turkey buzzards that had been slamming into the building for a time had disappeared.

Annette scanned the glassy facade of the executive suite two stories above where the buzzards had attacked so strangely the week before. A disturbingly prescient sign before launch, their absence now suggested, perhaps, a good turn.

The morning looked clear and sunny from her vantage. She had been at work since three A.M., but there were some here who hadn't taken a break for more than twelve hours, since five the previous afternoon.

It no longer felt quite so odd to be without her partners, Steve Gurney and Dan Rittman, even though they were the ones with whom she had spent most of her hours building motherboards and testing electronics for the past two and a half years. She and Gurney had practiced for hours on end in the lab upstairs conducting the same emergency maneuvers they now used to save the troubled satellites. Together they had built pieces of the spacecraft that would send and accept commands via what they called the "dirty pipe," from a gateway Earth station to the subscriber receiver aboard a satellite. Dan had left before the testing program ended, but she had completed the firecode reset operation with Gurney only a few weeks ago.

"Here you go, Annette," Gurney would say, "I'll be FM-1 at 200 degrees, you're FM-2 at 180 degrees. Here we go. . . ."

It was like play, at times. Innocent fun.

For a few weeks she thought one of them, either Dan or Gurney, should have been at the launch with her, at the gray consoles, digging through spacecraft transcripts as thousands of bit packets flowed in from low-Earth orbit every hour and a half or so. They had enjoyed each other's company despite everything else. What she missed most was their sense of humor.

For more than two years they had carried on in their first serious jobs out of school, forging a genuinely satisfying partnership. Real engineers, they should have been here, echoing her inevitable paranoia, being her sounding board for difficult decisions.

"Is there any other satellite around this small that works on negative control alarms?"

"No."

"Yes, there is."

Everyone sounded so very businesslike now, these colleagues who remained—some of whom she hardly knew at all—as they searched for what caused the spacecrafts' disabilities. Of course, under the eyes of Orbital's executives, they had to be.

Dan had moved back to Indiana. Gurney had left mentally weeks before he actually quit. While Annette was preparing for even greater responsibilities in the company by attending evening classes at George Mason, Gurney simply seemed to give up on the project entirely. At night he drifted off to shoot pool. During the day he disappeared more frequently from the lab to play Doom on one of the technicians' PCs with a half-eaten box of Jujubes at his side, until she tracked him down and forced him to talk to her. Those absurdist conversations, even in the final days together, lifted her spirits.

"When's the launch?"

"April Fool's Day."

"How appropriate."

"It'll be like: 'We launched this morning!'

'No we didn't.'

'Yes, we did!'

'No! What day is it?'

'April Fools!'

'Asshole.'

'I'm not joking.'"

It was his sense of humor she missed the most. But then everybody who once had partners, she imagined, missed someone now.

Sitting at the table, flipping through her own stiff stack of paper, the numbers seemed to float above the pages. Somewhere among the mush of telemetry points and health messages would be an answer to their dilemmas.

Sometimes the digits lined up like lucky numbers at the bottom of their fortune cookies—7-1-3-3-0-6-4-09—that always accompanied the only good news they received by dinnertime: "You will soon be relieved of a difficult burden" or "Happy days are here again." Privately, they joked that they might again see test incident 666, the famous anomaly that occurred at the launch site when someone inadvertently pressed the F-2 function key and sent separation commands to the spacecraft while it was still nestled in the rocket's fairing.

But everyone was somber today.

For a half-hour Annette and her colleagues reviewed the data and practiced a script they would use for sending the next firecode reset to satellites FM-1 and FM-2.

"Oh, one more thing," Annette said as the meeting started to break up and people returned to their positions in the control room or upstairs in the lab. "What's David Thompson saying about us?"

Around the corner the president of the company was seated at the glass perimeter, staring glumly at the projection of the Earth, like a big-time loser at the horse track.

"Why don't you ask him yourself?" one of the techs answered.

A few minutes later Annette leaned into the console, watching the edge of the moving circle drift slowly north from Mexico through the Midwest until it touched the western edge of New York State. Three little beeps sounded, indicating that the first satellite had connected with the Earth station in Arcade. She took a breath, then typed in the passwords before sending the firecode that everyone hoped would finally reset the satellite.

I_AM_DESPERATE

David Thompson couldn't see that much. At least the engineers surrounding her in the control room still appreciated the irony. They shared that one remaining play for humor. Although the goofy passwords were a vestige of times past, they still reminded everyone in the room of the team's cleverness, spunk, and independence. No matter how much he now regarded those once-estimable traits as a reflection of immaturity or youthful intelligence that had come into the company and become seriously flawed or woefully misspent, the fact was, a few members of the original team still sat at the controls.

They—the engineers and their satellites—were not dead yet.

For two weeks the team worked twenty-four-hour shifts. The most experienced engineers from other projects came in to troubleshoot, but for the most part Steffy's engineers did it all themselves: managing the satellites at the gray consoles; conducting dozens of simulations on the EDU spacecraft in the high bay; mimicking the performance troubles in orbit.

David Thompson vowed to leave them alone, but sometimes late at night he would sneak down from the fifth floor and stand silently outside the control center with a set of binoculars pressed to his eyes, reading the blue monitors as telemetry trickled in from the passing spacecraft.

Inside the control center, phones rang constantly. An accelerating loss of time and Murphy's law conspired against them. Beyond their protected realms—the control center, the high bay, the lab—everything was irrelevant, white noise, interference, caustic intramurals.

Fifteen days after launch, two A.M.:

"Okay, everybody, we're having a pass in, like, five seconds."

"Quick, Grace, exit this window, go into this file—here."

"What happened to the transmitter?"

"We still don't know."

"FM-1's over Arcade!"

"We don't have a connect. . . . Hey, Kim, where's our connect message? Where's our data?"

"We're not getting back-orbit data."

"No telemetry yet? What do you see on the solar array drive?"

"Zero."

"Battery charge current?"

"Nobody has telemetry."

"Nothing from the spacecraft?"

"Somebody call the high bay and tell them this trick's definitely not going to work."

"Annette, call the EDU lab. Ask if there's some other way to do a reset."

"Annette, Steffy's on the phone."

"Tell him we couldn't get enough RAM to allocate for these tables. . . . We don't have enough memory allocation. We need more memory allocation."

"Dave, we're not getting enough memory after we reset."

"There's not enough data."

"Tell him we're not getting any data."

"And we're not getting telemetry."

They heard no more laughter from the marketing department, no more flowers came in on delivery trucks from customers. Even the news media quit calling. Engineers on other projects finally speculated about what had doomed them. Orbcomm, they said, had been a product of "groupthink" without enough specialized, expert oversight. It gave way to its own hubris, to a lot of smart people who thought they could defy physics and live without margins. During the day, engineers from other programs going up and down the elevator would stop briefly at the third floor to see whether the satellites had spoken yet, and wondered out loud whether anyone would still have a job at the end of the year.

People from Alan Parker's group passed outside the center mumbling to each other: "If that was my team, I'd fire every damn one of them." Bottles of champagne Alan purchased for the celebration chilled inside a refrigerator down the hallway. Whenever satellite engineers completed a shift, they would head upstairs to their offices, close the blinds, and shut the door. They knew what was being said, even as they stayed sequestered. You could find some of them, hours later, curled in a fetal position beneath their desks.

And yet they had created satellites to survive disaster. While executives set up what they called a "war room" and cast about for solutions based on their years of aerospace experience, Steffy's engineers met privately at odd hours and methodically sketched out the entire satellite's software and hardware systems to ask, theoretically, what could have gone wrong.

The satellites had to live. They began their adventure on an umbilical and now traveled specific paths that allowed them only certain degrees of freedom. Certainly they spewed and spun, tumbled and outgassed, suffered bugs, skewed, required regular housekeeping, and experienced jitter. But that was normal. In time they would adapt independently to their environment and to their mistakes. The satellites could compensate for wild swings in temperature and odd changes in the physical environment, to solar winds and solar flares, radiation and magnetic fields, galactic noise, natural and terrestrial interference. The engineers knew the spacecraft would survive long enough to give them a chance to respond.

One thing they knew immediately: Their lack of margins did make Orbcomm spacecraft substantially different from other satellites. Other satellites, not driven by commercial pressures, were extensively shielded and protected from the natural environment of space. While government spacecraft bulged with redundant systems, radiation-hardened parts, and a large assortment of heaters, gyroscopes, thermal blankets, and propulsion tanks, Orbcomm relied more on the simple, weightless, and efficient intelligence of software.

Something foul, they suspected, had occurred in the code. Experimenting in the lab, they considered all the margins and compared their analysis with actual telemetry. Finally, they identified the problem in a single bit flip: The logic state of one of the microprocessors had randomly changed from 1 to 0 in binary memory, making programmed instructions aboard the satellite go haywire.

During one of its initial orbits, they realized, the satellites passed over an odd stretch of Earth off the Brazilian coast, where a peculiar geological formation created a natural disturbance in the Earth's magnetic fields. Known as the South Atlantic Anomaly, that one region of low-Earth orbit forced the Van Allen radiation belt to extend lower toward the Earth's surface, causing an unusually severe distortion in the environment. When the satellites crossed over South America, the intensity of the disturbance caused a bit flip in the satellites' RAM, knocked out communications, and made it impossible to reset the satellites' systems through RF channels commanded from the control center.

Dave Steffy had known about the satellites' susceptibility over the South Atlantic, but early in the project he was told that he could cut costs considerably by ordering commercial parts rather than radiation-hardened parts. An analysis of commercial processors, he was told, had concluded that latch-ups due to the anomaly would be extremely rare, occurring, theoretically, only once about every fifteen years. What he didn't know was that because of the expense of conducting detailed analysis on a large assortment of parts, the sample size of the tests amounted only to a small handful of processors.

For a while the team tried to convince the air force to send the satellites bad positioning data from its own GPS constellation. The idea was, if the satellites no longer knew where they were, then the attitude-control system would quit working, the spacecraft would flip upside down, cock the solar arrays away from the sun, and eventually discharge the batteries. At a certain level of discharge, the existing software would command the spacecraft to enter what was called Phoenix Mode. In Phoenix Mode the flight computer would automatically reset all the boxes, and reconfigure the software to amend the bit flips that confounded them.

But the air force refused. The satellites continued to orbit for a month in silence.

The only way left to save themselves was to improvise.

One day in mid-May I came to work and found Mark Krebs sitting outside the control center, chatting casually with his colleagues.

"So I see our next rocket project made *USA Today*," he said.

"Yeah, it was in *Av Week*, too," one of them said.

"And by the time it makes *The New York Times*, it'll be a prototype capable of interplanetary travel and establishing a colony on Mars. The usual hype."

But they were laughing, and inside the control center there was only one engineer at the consoles. They looked relaxed. They said John Stolte had left for the driving range to hit a bucket of balls. Eric Copeland had disappeared. Annette came around the corner with a big smile on her face.

"No more excuses," she said.

They were, in fact, tired but exuberant.

Apparently, early in the mission, the other spacecraft that launched with Orbcomm—the Microlab satellite containing the GPS and Lightning Sat experiments—had suddenly flipped upside down after someone uploaded bad positioning information. Called in to investigate, Krebs helped fix the problem and set the satellite on the right course again, but in the process he remembered that Orbcomm's attitude-control software suffered from one disastrous bug when it got positioning data, saying the spacecraft was flying beneath the Earth. Taking that one false ephemeride, the attitude-control system performed a calculation that divided by zero, and the control system crashed. Krebs knew the software in the attitude-control box would essentially freeze—not entirely, but in a convenient way that left all the health and maintenance functions active and fooled the flight computer into thinking everything was fine. In the meantime the satellite would drift out of position, the solar arrays would diverge from sunlight, the spacecraft would slowly lose power, and then it would enter Phoenix mode.

With one discrete uplink command the night before, Krebs had sent the satellites on an awkward tumble. Over a period of several orbits, the satellites lost power, entered Phoenix mode, and the software responded.

The reset happened immediately.

The satellites had begun to speak.

The project ended with Steffy's hip young team still jamming in the cybersphere. The Sinewave Sessions dragged on a few more weeks. But by the end of June, with the satellites' in-orbit checkout nearly complete and the team anxious to turn the keys over to Teleglobe, David Thompson sensed an exodus. He quickly called a meeting and pleaded with his engineers not to leave the company. He begged them to join a new team of veterans at the former Fairchild space company in Germantown, which was throttling up for production.

All they wanted to hear was that they had done the job, that B minus was good enough, and that he would not force them to move to Germantown. As he talked, the engineers listened carefully and skeptically.

"When we set out almost four years ago in the fall of 1991," Thompson said, "the goal was to be the first operational global messaging service. And we are first. We have achieved that goal. Since April we have gotten the satellites stabilized, and we have successfully sent messages. The satellites have pretty much completed acceptance testing, and the early service demonstrations are positive.

"If we look at the full scope of that undertaking, it's clearly a pioneering accomplishment. It was more than building satellites, more than launching them into orbit, more than setting up the ground structures to communicate. Your achievement with FM-1 and -2 ranks, in my mind, with the first Pegasus launch, except that your final push wasn't concentrated in a two-hour period—it went on for weeks. The magnitude of that accomplishment and its importance to the company is, arguably, even greater than the first Pegasus.

"You may think I was stupid in 1991 when I said a year and a half was not an unreasonable schedule. As things developed, we just didn't have enough people and we asked many of you to sustain the program a lot longer than we should have. We went through a number of crises, and some of those were, frankly, blown out of proportion. As you know, we have been under a terrific amount of external scrutiny.

"So we're well ahead of anybody else. But the satellites are far from perfect. We're still not there. The next phase, which entails improvements in our management and a thorough redesign of the satellites, will lead to a full production program at Germantown. With the investment from Teleglobe, we will build at least twenty-four more satellites, and we'll begin ramping up on that very quickly."

The engineers, standing around him in the conference room silently like a small congregation at vespers, expected more. They waited for him to conclude with a proper offer: more money, a realistic schedule, a promise not to move them to Germantown.

"I'd like to see you finish the job," he said. "I'd like to see the team reassemble. And I want to make that job attractive to you."

Eighty percent of the Orbcomm satellite design did not exist on paper. What Thompson and everyone else knew was that the details of the satellite design, its flaws, and its potential improvements existed mostly in the minds of its engineers.

After rushing their satellites to launch, seeing them pork-chopped in orbit, and then having to embrace their imperfections, ignobly, as a B-minus job and at the same time release them to perpetuity—well, it was still too much for any one of the engineers to stomach.

Trust me, Thompson seemed to say.

But they weren't sure that what he wanted them to do next would be a punishment or a privilege.

Stymied by their sudden ambivalence to imperfect objects and an imperfect world, among those who remained, Mark Krebs, Tony Robinson, Grace Chang, Kim Kubota, Kirk Stromberg, Annette Mirantes, Paul Stach, even John Stolte and Dave Steffy—Freshouts, Wolf Pups, and Lifers—no one knew exactly what to say.

They needed a long vacation to think, and then maybe they would consider an offer.

A few weeks later Teleglobe committed $70 million and vowed to support the business as a full partner. Steffy took over satellite production in Germantown, and Stolte joined him, as did Krebs and Robinson, Stromberg and Kubota.

I saw them for the last time together in August, when I dropped by to attend the first meeting on the satellite redesign and to sit in on an executive staff meeting. In Congress, it was reported, legislators would soon hear an upcoming budget proposal to take another $8 billion out of NASA over the next three years. Orbital's lobbyists were encouraging the White House to call for a more restrictive policy regarding the influx of Soviet missiles onto the world's commercial launch vehicle markets. Orbital's new Magellan subsidiary had opened a production facility in Mexico and predicted it would eventually build commercial GPS receivers and sell them for less than $100 a piece—the first truly mass-marketed space product. The commercial business at Orbital—and for space throughout the United States—was on the verge of takeoff.

At five o'clock, though, I ran into Kim Kubota on the fourth floor. She had new glasses and a new haircut. Grace was singing in the hallways still, though in a distinctively minor key. Kirk had just returned from a vacation in Hawaii. Steffy, Stolte, the others—they were gone.

"I'm going to leave and go play volleyball," Kim said.

Tony Robinson glanced out of his office at her, then looked at his watch, too.

"Quitting time!" he said.

"Tony! Not you!" Kim said, laughing.

"What d'you mean, not me? The satellites work. The stock's up. We've got the Teleglobe investment. And the last three years are a blur. I deserve a life again!"

Kim held a gym bag in hand. "I know, I can't remember a thing." She sighed.

"It's a different company," he said.

She agreed: "Everything's changed."

They left together, walking out into a still warm and sunny afternoon, oblivious to the darker regions overhead. The rapid pulse of satellites, speaking silently through the atmosphere, had at last relinquished them from the trance.

Engineering Savvy Called Key to Orbcomm Recovery

By Warren Ferster
Space News Staff Writer

WASHINGTON—Orbital Sciences Corp.'s ability to correct software problems in two orbiting spacecraft reflects an engineering savvy that rivals must duplicate to be successful in today's market, industry analysts said.

The company's May 22 announcement that it had made progress fixing problems with its Orbcomm data communications satellite will also help restore investor confidence in Orbital's plan to deploy a 26-spacecraft global messaging service by 1997, these analysts said. Orbcomm 1 is one of the two spacecraft launched last month to test and demonstrate the Orbcomm system. Engineers with the Dulles, Va., company have restored partial function to the subsystem on Orbcomm that allows individual subscribers to communicate.

"By recovering the satellites, Orbcomm has increased its credibility with investors and the data communications community," said Mary Ann Elliott, president and chief executive officer of Arrowhead Space and Telecommunications Inc., a Falls Church–based research and engineering firm specializing in satellite communications.

A command sent May 18 to Orbcomm 1 corrected a software blockage that was preventing the spacecraft from receiving transmissions from user terminals on the ground. The command reordered the sequence under which the receiver is designed to function, allowing it to operate without being first cued by the subscriber transmitter.

—*Space News*
May 29, 1995

"The reset was going to happen. I thought it might take years, like maybe in 1998 we would finally get it back, and you never know what would have broken between now and then."

—John Tandler
Orbcomm engineer
June 1995

"Not so long ago, the suits might not have even bothered to laugh at the dreamers. They would have simply dismissed them with a clubby smirk. No more. Instead, the big firms are starting to copy the start-ups. Many huge firms—Lockheed Martin, Boeing, and McDonnell Douglas, for instance—now have incubator 'commercial' divisions. Lockheed recently launched a new rocket design in only 27 months—warp speed by NASA standards."

—Heather Miller
Wired 4.09

"I would summarize the results of FM-1 and FM-2 by saying that while we are not as good as we sometimes think we are, we are still better than everybody else. Relative to what we had to achieve to keep Orbcomm moving forward, and relative to what others have thus far achieved in similar pursuits, I believe that our first small-satellite communications endeavor has been quite successful indeed."

—Bruce Ferguson, Orbital COO
Internal memo to David Thompson
February 15, 1996

"It is time to look beyond the adventure of exploring space. It is time to prove that space is a good investment and worth the risk. This is the new adventure."

—American Astronautical Society
National conference
Fall 1996

"Yet with so many ventures racing to embrace an uncertain marketplace, competition is going to be brutal. The only real question: 'Who is going to be the one making money?' asks Hughes Electronics Corp. Chairman Michael Armstrong. 'Unless you're first, you won't make much.' Steve D. Dorfman, who runs Hughes's Telecommunications & Space Co. is more direct: 'There's going to be a bloodbath,' he says."

—*Business Week*
January 27, 1997

Orbital Sciences to Buy CTA Satellite Division

By Martha M. Hamilton
Washington Post *Staff Writer*

Orbital Sciences Corp. yesterday announced a deal to pay $37 million in cash to buy the satellite manufacturing and communications division of CTA, Inc., of Rockville.

The purchase, which needs regulatory approval, would expand Orbital's existing business of making small, low-orbit satellites and take the company into a new one: production of larger communications satellites that orbit farther out in space.

"I think it is an excellent strategic acquisition," said Paul Nisbet, space analyst at JSA Research, Inc., in Newport, RI. "It fits beautifully. Orbital is buying its biggest competitor in the small satellite arena. It's now the preeminent small-satellite producer in the world."

Orbital, which has 3,200 employees, half of them in the Washington area, is buying a division that had $80 million in sales last year and employs 300 people.

Orbital spokesman Barron Beneski said that all 300 would continue to work at a facility in McLean but eventually would "migrate" to Orbital facilities in Germantown and Sterling.

—*The Washington Post*
July 15, 1997

Orbital Subsidiary Introduces First GPS Rx Priced Under $100

From the company that introduced the world's first handheld Global Positioning System satellite navigator comes the first new portable GPS receiver priced under $100.

The "GPS Pioneer" personal navigator, from Orbital's Magellan subsidiary, is now available at over 10,000 retail outlets in the United States. Designed to enhance any outdoor adventure, the GPS Pioneer uses the same advanced satellite technology that guides troops, navigates oil tankers, steers jets, and helps explorers on ascents of Mount Everest.

—Orbital Sciences Corp.
January 1998

"Orbital Sciences Corporation (ORBI NASDAQ) today announced its full-year 1997 financial results, reporting annual revenues of $605,975,000, an increase of 31 percent over 1996 annual revenues. The company also reported net income of $23,005,000 for the years, up 45 percent compared to net income of $15,907,000 in 1996."

—Orbital Sciences Corporation
February 5, 1998

Epilogue

David Thompson recently told an interviewer he had purchased twenty copies of a book by Warren Bennis titled *Organizing Genius*, which investigates the characteristics of successful teams, such as the one that saved Apollo 13 from disaster.

"Why," he wondered out loud, "does one group click and produce far more than the ordinary, while the other one never makes it?"

The more relevant question seemed to be: Why does he care?

In late September 1998, a Pegasus rocket lobbed eight more Orbcomm satellites into the sky, bringing the global messaging system up to full service with a total of twenty-eight spacecraft. Although it took three years from the launch of FM-1 and FM-2 to finish the job, the redesigned satellites now cover the globe. Orbital remains a step ahead of the competition.

In 1997 the company carried out twenty-three satellite and rocket launches, with a 100 percent success rate. When 1998 began, the company's stock hit an all-time high. Orbital reported record revenues of $605 million in 1997, began 1998 with a total order backlog of $2.9 billion, and received $1 billion in new orders during the first half of that year. After strenuous efforts to rebuild its rocket and satellite teams, to merge and acquire more companies, and to introduce new production techniques, Orbital's executives can reasonably expect the gamut of Orbital space products to generate more than $1 billion in revenues before the year 2000.

Continuing on a brisk pace toward growth and expansion, Orbital recently became one of the ten largest satellite-related companies in North America, with 1998 earnings estimated at $750 million. True to the founders' goal to build a vertically integrated space company offering end-to-end services, David Thompson can boast at last that his company designs, manufactures, operates, and markets total space systems—satellites, satellite services, launch vehicles, sensors, electronics, Earth stations, software, satellite navigation and communication products, and Earth-imaging products.

Also, true to the company's purpose to sell products that will bring the resources of space to consumers around the globe, Orbital recently introduced the world's first portable GPS receiver priced under $100 and opened a new market for "intelligent" transportation with its PathMaster vehicle navigation system, now being used on snowplows, paint, and herbicide trucks for public works departments in Colorado, Maryland, Iowa, and Oregon. Orbcomm has signed eleven service providers, covering ninety countries (approximately 1.2 billion people) around the world, making the promise of an inexpensive Dick Tracy personal communicator a reasonable possibility by the twenty-first century.

In many ways he has proven that the company can employ engineers, create teams, and build space products far beyond the ordinary.

So why does David Thompson worry about *organizing genius*?

Simple: The competitive nature of personal satellite markets has become particularly severe. During the second week of September 1998, Loral's Globalstar system lost twelve satellites in its forty-eight-satellite constellation when the Ukrainian rocket carrying them exploded en route to orbit and left the $190 million cargo raining down over Siberia. With its mobile phone service suddenly delayed until the end of 1999, Globalstar saw a large percentage of its Wall Street investors pull out. Executives responsible for the $2.6 billion constellation were left scrambling to find a new launch vehicle and make plans to operate the system with only thirty-six satellites.

At the same time , Globalstar's chief rival, Iridium, was struggling to cope with a number of in-orbit failures in its sixty-six-satellite system. Just as the company began launching spares, Iridium announced it would delay the start-up of its worldwide phone service until November of 1998. With the price for one of its phones still retailing at more than $3,000, executives said they will give away a number of phones, at least initially, because they cannot guarantee the service will work successfully from the start.

Orbital slots that David Thompson anticipated more than ten years ago have filled rapidly. So many companies have entered the fray that the question about who will succeed and who will fail continues to haunt the business. Industry experts have concluded that the world needs only two or three LEO systems, and investment advisers are predicting a bloodbath among the losers. Even if only two or three survive the battle for control, the first-to-market strategy will likely help Orbcomm break through the inaugural years.

It's said that the technology improves so quickly, one day Orbcomm is number one in the race, the next it is number ten. One day Iridium commands the world, the next day it's Globalstar, the next, Teledesic. For instance, Teledesic's engineers have redesigned their system to use *only* 288 satellites rather than more than 600, which were originally planned just three years ago. And Iridium's engineers have also begun to redesign their satellites, even as the originals begin coming into service, so they can upgrade the constellation before any new competitor beats them to it.

At this moment, too, Orbital has several projects under way to replace existing Orbcomm satellites with faster, cheaper, better spacecraft. It recently won a $260 million contract with a company called VisionStar to build two geostationary satellites for high-speed Internet connections and broadcasting, and the company has announced plans to build a twelve-satellite constellation for mobile phone service by 2001.

The aerospace business today looks more and more like the personal computer industry of the early 1980s, where this afternoon's design needs upgrading by tomorrow morning. A treacherous business, it remains enormously expensive, extremely risky, as ferocious as a shoot-out on the Klondike, and potentially as rewarding as a riverbed glittering with gold.

But there is also another reason that David Thompson cares so much about "organizing genius."

Recalling Orbcomm's early years, he compares the time to a prolonged battle, a stretch of tactical defeats that resulted in one lasting, strategic victory. But it is also apparent that the company was swept up in the beginnings of a revolution that will not soon end. Commercial space ventures—once the oxymoron of the aerospace business—now attract substantial capital resources from the nation's most accomplished aerospace businesses. The LEO frontier represents only one battlefield in the revolution. The larger war rages on.

Revolutions are messy, Thompson will say, and the story of the first years of the Orbcomm project proves that. Looking back, the evidence is enough to make him squirm.

For example, a financial analysis in 1996 showed that Orbital spent $26.2 million for the FM-1 and FM-2 spacecraft, and another $12.5 million for their first Pegasus rocket ride—more than anyone anticipated, but not exorbitant, considering the challenges and complications. However, further review also confirmed that the satellite builders' complaints were genuine. Orbital's executives seriously underestimated the technical challenges in creating high-performance spacecraft. The repeated launch delays seriously undermined the company's credibility with communications suppliers, resellers, foreign affiliates, customers, and investors.

What seems clear now is that while Dave Steffy's team built the first satellites, he and his engineers also created the basic infrastructure so the company could eventually become a significant spacecraft manufacturer. They set engineering standards, acquired test equipment, developed a stockroom, set configuration-control techniques, established a sensible quality-control team, hired contractors, and built enduring relationships with vendors. The young

Freshouts broke new ground in on-orbit spacecraft position management, in antenna design and deployment techniques, and in overall system architecture.

Having survived the battle, Orbital now has a satellite-development and manufacturing business bigger than the entire company was in 1992—a third of a billion-dollar business. The hard work of the first Orbcomm team has led to investments of more than $500 million since 1995, far more than the $100 million Thompson initially thought he needed to deploy a constellation.

Typically, Thompson wonders what might have happened if he had been smarter. But then, Orbital was not TRW. His engineers hadn't built four hundred satellites over the past thirty years.

"We were just a little company," Thompson says. "So it was messy. It was our Bunker Hill. I look back now, and it's like these ghosts—you can see their faces, the bandages wrapped around their heads, and you know some of them won't return for the next fight. Sometimes when I think about that, it seems like we didn't do anything right. But then, okay, I explain it by looking at how the Continentals finally stood up to the British army. They lost the battle—the British had three-to-one numerical superiority and they had their boats parked down on the Charles River and shot 'em up with cannons—and yet the Continentals held them off for ten days and they didn't run away, and their fight emboldened a spirit to win the war. For the next several years, everyone thought, "Bunker Hill," and fought like hell.

"That was us. We experienced a whole series of errors and retreats—and sometimes we decided we just didn't want to fight at all that day. But then you take the long view and ask, so how did the war end? What did that battle mean to us?

"In the American Revolution it would have been nice, I guess, if the Continentals had had this well-trained professional army, but that was impossible because they didn't even have a country yet. We were living through a revolution, in a sense, too—only our revolution was happening inside the company and, to a much greater extent, inside a dozen other companies who were fighting the same war in this revolution of blunders and great leaps of technical genius."

Despite their success, Dave Steffy's small team of engineers who eventually developed critical building blocks for the world's first commercial constellation have now vanished. A few stayed with the company, but most left without waiting to enjoy the final reward.

I do, however, still hear from them. Mark Krebs, Morgan Jones, Steve Gurney, and a few of the others occasionally write to say where they are now. Some, like Gurney, have gone on to graduate school to earn MBAs and enter other industries. Some, like Morgan Jones and Rob Denton, have started their own small businesses outside aerospace. Mike Dobbles moved to California to work in the software industry. Tony Robinson took a job with Teledesic and is now designing Bill Gates's first satellites.

Of course, the passage of time gives them all a new appreciation for the Orbcomm accomplishment. Tony notes that his new work has been so conventional, by comparison, that for the first two years he was only building "paper satellites" and had not been able to get his hands on a single piece of hardware. Krebs, who recently left Orbital to start his own business, still expresses gratitude for his years on the team. He is particularly expressive when he recalls how he learned that the force of human willpower can overcome the chaos of space and effortlessly defy laws of "gravity, probability, and entropy." He wrote me recently to say he actually left the company with some regret, noting: "I will miss this business more than I can explain to you, and probably more than I can imagine myself."

As far as I can tell, they are all quite happy, no matter what industry they have now chosen or what challenges they now pursue. The once idealistic and eager engineers in David Thompson's first class of Freshouts and Wolf Pups remain optimistic and confident, stronger by measure for their prolonged trial. Many of them report they are again chasing the next Big Cookies of a new millennium, just as legions of engineers and managers at Orbital Sciences continue to do, eagerly staking claims in commercial frontiers from development to production to launch to service.

What they will create next, who will benefit from their new technologies, what will happen next, only time will tell. But one thing is clear: As this decade ends and the next begins, the new space age already bears marks of genius, consecrated by the sweat and toil of dozens of anonymous men and women in secluded labs where heroic efforts are often barely sufficient for the day. They join a thrillingly enormous battle, carrying bright banners into the rugged territory of commercial space.

Gary Dorsey
Baltimore, Maryland
January 1999

Cast of Characters

Tony Austin: satellite quality assurance manager
Gregg Burgess: satellite principal systems engineer
Stan Ballas: Orbcomm gateway Earth station
Frank Bellinger: satellite integration and testing manager
Mike Carpenter: satellite communications, senior engineer
Grace Chang: satellite attitude control engineer
Bill Cleary: satellite software engineer
Bill Coffman: Microlab project manager
Nancy Cohen: satellite communications engineer
Eric Copeland: satellite communications, senior engineer
Shawn Curtin: satellite senior designer
Rob Denton: satellite battery control and power system, electrical engineer
Mike Dobbles: satellite propulsion, associate mechanical engineer
Mickey Doherty: satellite technician
Mark Dreher: Orbcomm marketing director
Bruce Ferguson: Orbital chief operating officer
David Figueroa: Orbcomm communications engineer
Steve Gurney: satellite software and ground control engineer
Gordon Hardman: satellite communications engineer (Boulder)
Morgan Jones: satellite operating system, associate systems engineer
Jan King: Boulder division director
Mark Krebs: attitude control, senior engineer
Kim Kubota: satellite testing, senior systems engineer
Bob Lindberg: Apex satellite project manager
Paul Locke: Orbcomm, chief engineer
Bob Lovell: Orbital Space Systems Division president
Steve Mazur: Orbcomm systems and network manager

John Mehoves: Space Systems Division executive vice president

Annette Mirantes: satellite software engineer

Alan Murray: satellite technician

Alan Parker: Orbcomm president

Rob Phillip: satellite operating system engineer

Dan Rittman: satellite gateway communications, electrical engineer

Tony Robinson: satellite mechanical, senior engineer

David Schoen: Orbcomm director of engineering

Dave Steffy: Orbcomm satellite project manager

Paul Stach: satellite testing, associate engineer

John Stolte: Orbcomm satellite assistant project manager

Kirk Stromberg: satellite testing, systems engineer

Don Thoma: Orbcomm marketing

David Thompson: Orbital president and chief executive officer

J. R. Thompson: Orbital executive vice president

Jim Utter: Orbital vice president for business operations

Bibliography

Bainbridge, William. *The Spaceflight/Revolution: A Sociological Study.* New York: Krieger Publishing Co., 1976.

Barter, Neville J. *TRW Space Data.* Redondo Beach, CA: S&TG Communications, 1992.

Bertstein, Jeremy *Experiencing Science.* New York: Basic Books, 1978.

Burrows, William *Deep Black.* New York: Random House, 1986.

———. *Exploring Space.* New York: Random House, 1990.

Bush, Vannevar. *Modern Arms and Free Men.* New York: Greenwood Publishing Group, 1949.

Byerly, Jr., Radford, ed. *Space Policy Alternatives.* Boulder, CO: Westview Press, 1992.

Chander, Romesh. *Planning for Satellite Broadcasting: The Indian Instructional Television Experiment.* Paris: UNESCO Press, 1976.

Chetty, P. R. K. *Satellite Technology and its Applications.* Blue Ridge Summit, PA: Tab Professional and Reference Books, 1991.

Clarke, Arthur C. *The Exploration of Space.* New York: Harper & Brothers, 1951.

———. *Ascent to Orbit: A Scientific Autobiography (The Technical Writings of Arthur C. Clarke).* New York: John Wiley & Sons, 1984.

———*Voice from Across the Sea.* New York: Harper & Row, 1974.

———*The View from Serendip.* New York: Random House, 1977.

Demac, Donna, ed. *Tracing New Orbits.* New York: Columbia University Press, 1986.

De Sola Pool, Ithiel. *Technologies Without Boundaries.* Cambridge: Harvard University Press, 1990.

Finch, Edward Ridley, and Amanda Lee Moore. *Astrobusiness: A Guide to Commerce and Law of Outer Space.* New York: Praeger Publishers, 1985.

Fraser, Ronald. *Once Round the Sun: The Story of the International Geophysical Year.* New York: MacMillan, 1958.

Green, Constance M., and Milton Lomask. *Vanguard: A History.* Washington: Smithsonian Institution Press, 1971.

Gunter, Paul. *The Satellite Spin-Off.* New York: Robert B. Luce, 1975.

Hall, Stephen. *Mapping the Next Millennium.* New York: Random House, 1992.

Hawke, David Freeman. *A History of Technology, 1776–1860.* New York: Harper & Row, 1988.

Howell, Jr., W. J. *World Broadcasting in the Age of the Satellite.* Norwood, NJ: Ablex Publishing, 1986.

Jaffe, Leonard. *Communications in Space.* New York: Holt, Rinehart and Winston, 1966.

Jasani, Bhupendra, ed. *Space Weapons and International Security.* London: Oxford University Press, 1987.

King-Hele, Desmond. *A Tapestry of Orbits.* London: Cambridge University Press, 1992.

Klass, P. J. *Secret Sentries in Space.* New York: Random House, 1979.

Lang, Daniel. *From Hiroshima to the Moon.* New York: Simon & Schuster, 1959.

Lent, John, and Gerald Sussman, eds. *Transnational Communications: Wiring the Third World.* Newbury Park, CA: Sage Publications, 1991.

Mack, Pamela. *Viewing the Earth.* Cambridge: MIT Press, 1990.

McDougall, Walter A. *. . . The Heavens and the Earth.* New York: Basic Books, 1985.

McLucas, John. *Space Commerce.* Cambridge: Harvard University Press, 1991.

Meindl, James, ed. *Brief Lessons in High Technology.* Stanford: Stanford Alumni Association, 1991.

O'Neill, Gerard K. *2081: A Hopeful View of the Human Future.* New York: Simon & Schuster, 1981.

Ordway, Frederick I., III, and Mitchell R. Sharpe. *The Rocket Team.* New York: Thomas Y. Crowell, 1979.

Pierce, John R. *The Beginnings of Satellite Communications.* San Francisco: San Francisco Press, 1968.

Pierce, John R., and A. Michael Noll. *Signals: The Science of Telecommunications.* New York: Scientific American Library, 1990.

Shane, Scott. *Dismantling Utopia: How Information Ended the Soviet Union.* Chicago: Ivan R. Dee, 1994.

Shternfeld, Ari. *Soviet Space Science: The Story of Artificial Satellites.* New York: Basic Books, 1959.

Stares, Paul B. *The Militarization of Space: U.S. Policy, 1945–1984.* Cornell University Press, 1985.

Sullivan, Walter. *Assault on the Unknown.* New York: McGraw-Hill, 1961.

Tedeschi, Anthony Michael. *Live Via Satellite.* Washington: Acropolis Books, 1989.

Weiner, Jerome. *Planet Earth.* New York: Bantam Books, 1986.

Index